Lecture Notes in Physics

New Series m: Monographs

The Editorial Policy for Monographs

The series Lecture Notes in Physics reports new developments in physical research and teaching - quickly, informally, and at a high level. The type of material considered for publication in the New Series m includes monographs presenting original research or new angles in a classical field. The timeliness of a manuscript is more important than its form, which may be preliminary or tentative. Manuscripts should be reasonably self-contained. They will often present not only results of the author(s) but also related work by other people and will provide sufficient motivation, examples, and applications.
The manuscripts or a detailed description thereof should be submitted either to one of the series editors or to the managing editor. The proposal is then carefully refereed. A final decision concerning publication can often only be made on the basis of the complete manuscript, but otherwise the editors will try to make a preliminary decision as definite as they can on the basis of the available information.
Manuscripts should be no less than 100 and preferably no more than 400 pages in length. Final manuscripts should preferably be in English, or possibly in French or German. They should include a table of contents and an informative introduction accessible also to readers not particularly familiar with the topic treated. Authors are free to use the material in other publications. However, if extensive use is made elsewhere, the publisher should be informed.
Authors receive jointly 50 complimentary copies of their book. They are entitled to purchase further copies of their book at a reduced rate. As a rule no reprints of individual contributions can be supplied. No royalty is paid on Lecture Notes in Physics volumes. Commitment to publish is made by letter of interest rather than by signing a formal contract. Springer-Verlag secures the copyright for each volume.

The Production Process

The books are hardbound, and quality paper appropriate to the needs of the author(s) is used. Publication time is about ten weeks. More than twenty years of experience guarantee authors the best possible service. To reach the goal of rapid publication at a low price the technique of photographic reproduction from a camera-ready manuscript was chosen. This process shifts the main responsibility for the technical quality considerably from the publisher to the author. We therefore urge all authors to observe very carefully our guidelines for the preparation of camera-ready manuscripts, which we will supply on request. This applies especially to the quality of figures and halftones submitted for publication. Figures should be submitted as originals or glossy prints, as very often Xerox copies are not suitable for reproduction. In addition, it might be useful to look at some of the volumes already published or, especially if some atypical text is planned, to write to the Physics Editorial Department of Springer-Verlag direct. This avoids mistakes and time-consuming correspondence during the production period.
As a special service, we offer free of charge LaTeX and TeX macro packages to format the text according to Springer-Verlag's quality requirements. We strongly recommend authors to make use of this offer, as the result will be a book of considerably improved technical quality. The typescript will be reduced in size (75% of the original). Therefore, for example, any writing within figures should not be smaller than 2.5 mm.
Manuscripts not meeting the technical standard of the series will have to be returned for improvement.
For further information please contact Springer-Verlag, Physics Editorial Department II, Tiergartenstrasse 17, W-6900 Heidelberg, FRG.

M. Klein A. Knauf

Classical Planar Scattering by Coulombic Potentials

Springer-Verlag Berlin Heidelberg GmbH

Authors

Markus Klein
Andreas Knauf
Technische Universität, Fachbereich 3 Mathematik, MA 7-2
Straße des 17. Juni 135, W-1000 Berlin 12, FRG

ISBN 978-3-662-13900-4 ISBN 978-3-540-47336-7 (eBook)
DOI 10.1007/978-3-540-47336-7

Originally published by Springer-Verlag Berlin Heidelberg New York in 1992
Softcover reprint of the hardcover 1st edition 1992

Typesetting: Camera ready by author/editor using the LATEX macro package from Springer-Verlag Berlin Heidelberg GmbH.
58/3140-543210 - Printed on acid-free paper

Contents

1. Introduction

Astronomy as well as molecular physics describe non-relativistic motion by an interaction of the same form: By Newton's respectively by Coulomb's potential.

But whereas the fundamental laws of motion thus have a simple form, the *n-body problem* withstood (for $n > 2$) all attempts of an explicit solution.

Indeed, the studies of Poincaré at the end of the last century lead to the conclusion that such an explicit solution should be impossible.

Poincaré himself opened a new epoch for rational mechanics by asking *qualitative* questions like the one about the stability of the solar system.

To a large extent, his work, which was critical for the formation of differential geometry and topology, was motivated by problems arising in the analysis of the n-body problem ([38], p. 183).

As it turned out, even by confining oneself to questions of qualitative nature, the general n-body problem could not be solved. Rather, simplified models were treated, like *planar* motion or the *restricted* 3-body problem, where the motion of a test particle did not influence the other two bodies.

In this work, we analyse qualitative aspects of the planar scattering of such a particle in the attracting fields of n fixed celestial bodies (resp. nuclei). We examine the dynamics generated by the Hamiltonian function

$$\hat{H}(\vec{q},\vec{p}) = \tfrac{1}{2}\vec{p}^{\,2} + V(\vec{q}), \tag{1.1}$$

where we allow the potential

$$V(\vec{q}) = \sum_{l=1}^{n} \frac{-Z_l}{|\vec{q} - \vec{s}_l|} + W(\vec{q}), \qquad Z_l > 0$$

to contain a smooth term W, in addition to the purely Coulombic sum. The only restriction on W consists in the assumption that for large distances V decays like a Coulomb potential, *i.e.* $V(\vec{q}) \sim -Z_\infty / |\vec{q}\,|$, where the asymptotic charge Z_∞ may be positive, negative or neutral.

The motivation for W comes from scattering by an n-atomic molecule, where the potential W is induced by the mean charge distribution of the bound electrons. In Fig. 1.1 we show some scattering orbits in the field of $n = 3$ Coulomb potentials.

Our research is based on methods from differential geometry and topology. In one sentence, our main result is the statement that for large energies, the

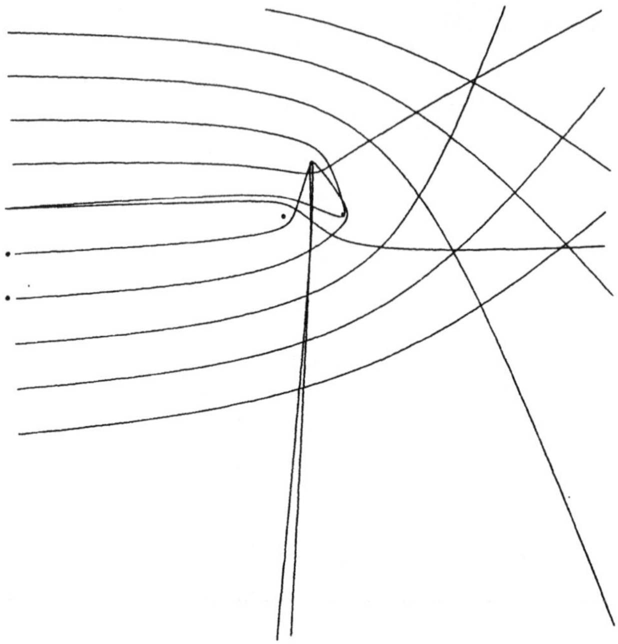

Fig. 1.1. Scattering trajectories for fixed initial angle and energy in the field of $n = 3$ centers

qualitative traits of the scattering process do not depend on the precise form of the Coulombic potential but only on the number n of centres.

For $W = 0$, the case $n = 1$ corresponds to Keplerian motion (or Rutherford scattering), whereas $n = 2$ is the two-centre problem which was solved by Jacobi. In the cases $n \geq 3$ one observes *irregular scattering.* In particular, the time delay diverges on a set of initial conditions which has a Cantor structure (see Figures 10.2 b), 10.3 a) and b)).

The term 'irregular scattering' has been coined only five years ago, but as it happens quite often, some ideas go back to the last century. It was Hadamard who investigated the form of geodesics on a special surface of negative curvature. This surface had three 'horns' going to infinity. Hadamard noticed in [15] that the geodesics which did not go to infinity formed a Cantor set. In §58 of his study [16] he re-stated that theorem for rather arbitrary negatively curved surfaces with 'horns'. Then, in §59, he asked:

> "Les circonstances que nous venons de rencontrer se retrouveront-elles dans d'autres problèmes de Mécanique? Se présenteront-elles, en particulier, dans l'étude des mouvements des corps célestes? C'est ce qu'on ne saurait affirmer. Il est probable, cependant, que

> les résultats obtenus dans ces cas difficiles seront analogues aux précédents, au moins par leur complexité." [1]

It is somewhat ironic that surfaces which are exactly of the type treated by Hadamard will be used in this study to show the presence of similar Cantor sets of bounded orbits in our planar n-centre problem.

In fact, the *bounded* orbits are the key to the understanding of irregular scattering. Their (un)stable manifolds extend to spatial infinity. Orbits on these manifolds are neither scattering nor bounded, but they influence the neighbouring scattering orbits and create their irregularity.

Beginning with the works of Eckhardt, Jung and others (see [9],[8]), since about 1986 a growing group of physicists is working on questions of classical irregular scattering. In the last years, the Quantum Mechanics of irregular scattering drew more and more attention.

The state of the art is nicely documented in the review articles of Eckhardt [8] (1988), Smilansky [41] (1989), and Tél [43] (1990), and we refer to the extensive references therein.

Below, we shall work out some *specifics* of the system (1.1). Two aspects distinguish our model from the majority of those analysed in the literature; on the one hand the attracting nature of the interaction, on the other hand its singular form (which however, contrary to the equally singular billard problems, leads to an energy dependent dynamics).

Our own interests in the model (1.1) has different roots. In Knauf [27], one of us considered the motion of a classical electron in a periodic potential with Coulombic singularities. It turned out that, for large energies, the motion of the particle in the planar crystal corresponded to a deterministic diffusion process. In [28] it was shown that the (ballistic) motion of a quantal electron in such a potential was retarded by quantum chaotic effects, and that its speed went to zero in the semiclassical limit $\hbar \searrow 0$.

In Klein [22, 23], and in Combes, Duclos, Klein and Seiler [7] the resonances of a quantum mechanical scattering process were related to classical bound states of positive energy. A certain virial condition, which implies (without being equivalent to) the absence of such orbits, was proved to lead to an $\hbar$-independent resonance free zone near the real axis in the complex energy plane. To the contrary, in the shape resonance problem, stable periodic orbits of the Hamiltonian vector field associated to a potential with a positive minimum lead to resonances whose imaginary part is exponentially small in $\hbar$ as $\hbar \searrow 0$.

We remark that, in contrast to these papers, a finer analysis of the semiclassical limit is based upon the work of Helffer and Sjöstrand [17], which uses microlocal analysis. This allows to explore in greater depth the consequences of the structure of classical motion for the semiclassical limit of Quantum Mechanics.

---

[1] "Can one find the preceding circumstances in other problems of mechanics? Are they present, in particular, in the analysis of the motions of celestial bodies? One could not be shure about that. Nevertheless, it is probable that the results obtained in these difficult cases are analogous to the preceding ones, at least with respect to their complexity."

This work, too, is related to quantum mechanical questions, which will be dealt with in [24]. We will mention connections to Quantum Mechanics in the general survey below.

To begin with, we describe in *Chap. 2*, Def. 2.1 the class of Coulombic potentials V to be considered.

One important and not merely technical aspect of our work is the *regularization* of collision orbits. In Prop. 2.3 this problem is solved by finding a completion P of phase space. P is a smooth four-dimensional manifold, and the Hamiltonian flow Φ^t on P is smooth, too. So the Coulomb 'singularity' turns out to be a smooth knot in phase space.

The *Møller transformations* $\Omega^\pm$ are the central object of classical scattering theory. Because of the long range of the Coulombic potential V one cannot define them by comparing the flow Φ^t with the free motion. Instead, we use the Kepler motion Φ^t_∞ generated by the point charge Z_∞ in the definition (2.22) of $\Omega^\pm$.

In the whole book, we make use of the so-called *virial identity*

$$\frac{d}{dt}(\vec{q}(t)\cdot\vec{p}(t)) = 2(E - V(\vec{q}(t))) - \vec{q}(t)\cdot\nabla V(\vec{q}(t)).$$

This identity is extremely useful since it implies the existence of an *interaction zone* in the configuration plane which, once left by the particle, cannot be reentered.

A consequence of the very existence of the Møller operators and their inverse (Propositions 2.6, 2.7) is *asymptotic completeness*, *i.e.* the fact that the Liouville measure of those phase space points, which are neither bound nor scattering states, is zero (Cor. 2.8).

In *Chap. 3* we introduce yet another regularization scheme. Using the so-called Jacobi metric, we translate the original Hamiltonian dynamics Φ^t for energy E into the equivalent problem of geodesic motion in that metric. In a second step, instead of considering geodesic motion in the original configuration plane, we define a Riemann surface $\mathbf{M}$ by

$$\mathbf{M} := \left\{ (q,Q) \in \mathbb{C}\times\mathbb{C} \,\middle|\, Q^2 = \prod_{l=1}^{n}(q - s_l) \right\}. \tag{1.2}$$

$\mathbf{M}$ covers the configuration plane twice, and the branch points of the covering are situated at the locations s_l of the nuclei.

Lifting the (singular) Jacobi metric to $\mathbf{M}$, one obtains a *smooth* Riemannian manifold $(\mathbf{M}, \mathbf{g}_E)$ (Prop. 3.1). Henceforth we analyse the geodesic motion on that surface in order to understand our original problem in scattering theory.

$\mathbf{M}$ is a surface with $[(n-1)/2]$ handles (Lemma 3.4). The Gaussian curvature K_E of the Jacobi metric $\mathbf{g}_E$ regulates the stability or instability of the geodesic flow Φ^t_E. $(\mathbf{M}, \mathbf{g}_E)$ is asymptotically flat, since by definition of V the force $-\nabla V(\vec{q})$ and its variation go to zero as $|\vec{q}| \to \infty$. The total curvature $\int_{\mathbf{M}} K_E(\vec{q}) d\mathbf{M}_E = -2\pi \cdot n$ is negative (Lemma 3.3). For purely Coulombic potentials (that is, $W = 0$), $K_E < 0$ for all positive energies. So the drawings in Fig. 3.1 of the Riemann surface preserve some aspects of the metric $\mathbf{g}_E$.

The fundamental group $\pi_1(\mathbf{M})$ forms a bridge between the topology of $\mathbf{M}$ and the closed geodesics on that surface. For n nuclei, $\pi_1(\mathbf{M})$ is a free group on $n-1$ generators (Lemma 3.5). In particular, it is non-abelian for $n \geq 3$.

For all energies E larger than the supremum of V, we find in *Chap. 4* closed geodesics on $(\mathbf{M}, \mathbf{g}_E)$. These geodesics project to closed trajectories of our original flow Φ^t, using the covering construction. For every non-trivial conjugacy class there exists a representative which is such a closed geodesic (Thm. 4.6).

As a technical tool, we minimize an energy functional (4.1) on the (infinite dimensional) Hilbert-manifold of loops in the surface $(\mathbf{M}, \mathbf{g}_E)$. Completeness of a Riemannian manifold carries over to completeness of its loop space (Prop. 4.1), and in our case, for non-trivial loops, the so-called Palais-Smale condition is met.

Negative Gaussian curvature leads to instability of the geodesics, but for a general Coulombic potential, K_E need not be negative everywhere. Nevertheless, in *Chap. 5* we find estimates which imply instability for large energies.

The idea behind these estimates is the following: for E large, trajectories which stay for a long time in the interaction zone near the nuclei must have close encounters with these nuclei again and again. But near these nuclei K_E is known to be negative, leading to instability.

In Prop. 5.1 we show that trajectories, which do not have encounters of distance $\mathcal{O}(E^{-1})$ with the nuclei, leave the interaction zone after a short time.

One problem in the very definition of orbital instability is the fact that all scattering orbits ultimately leave the interaction zone so that their Lyapunov exponent is zero. Yet we would like to describe the instability of the motion in the interaction zone in a manner which applies to the scattering orbits as well as to the semibounded and bounded orbits.

In that respect the *cone field* technique of Wojtkowski [48] turns out to be useful. Originally invented for analysing motion in compact manifolds, it easily generalises to our situation of noncompact energy shells $\Sigma_E = H^{-1}(E)$. In every plane perpendicular to the flow direction in the energy shell, we select a cone, which, roughly speaking, contains the unstable direction. If one is able to find such a cone field which is strictly invariant w.r.t. the positive time flow (*i.e.*, if the transported cones are strictly contained in the local cones), then the motion is unstable.

In Prop. 5.7 we show the existence of such a strictly invariant cone field for large energies. We are interested in the energy dependence of the scattering process. The concrete estimates given in Prop. 5.7 imply a divergence of the Lyapunov exponent of the bounded orbits like $\sqrt{E}\ln E$.

That kind of nontrivial, universal high energy behaviour is somewhat special to our model, because it depends critically on the fact that the Coulombic potential V is singular. For bounded potentials the high-energy scattering would be confined inside a cone around the forward direction of size $\mathcal{O}(E^{-1})$.

In *Chap. 6* we establish *symbolic dynamics* for the bounded geodesics on the surface $(\mathbf{M}, \mathbf{g}_E)$.

Since we do not want to loose any information on the energy dependence of the motion, we try to construct every object in a rather explicit way. Clearly, the starting point is the fundamental group $\pi_1(\mathbf{M})$ of our surface, and we find $n-1$ closed geodesics $\mathbf{c}_l$ which generate this group (Lemma 6.1). These geodesics project to trajectories in the configuration plane which start from the n-th singularity and then collide with the l-th singularity, bouncing back again.

We are enclined to think of a general bounded geodesic as a random walk on the fundamental group. To make this idea precise, we find in Lemma 6.2 $n-1$ geodesic segments $\mathbf{d}_l$ which are dual to the $\mathbf{c}_l$ in the sense of intersection numbers:

$$\#(\mathbf{d}_l, \mathbf{c}_k) = \delta_{lk}.$$

Then we erect *Poincaré sections* projecting to the $\mathbf{d}_l$. To find a symbolic description of the bounded geodesics, we consider a suspension flow over a bi-infinite sequence of symbols, the symbols numbering the Poincaré sections.

In Thm. 6.11 we show that for E large, a similar suspension flow is conjugate to the flow Φ^t, restricted to the set b_E of bound states of energy E. In particular we have just one bounded orbit for $n = 2$, bouncing back and forth between the nuclei, whereas for $n \geq 3$, the set b_E is locally of the form of a Cantor set times an interval.

The location of the semiclassical resonances in the two-centre problem will depend on the length of the closed orbit in the Jacobi metric, and on its Lyapunov exponent [24].

The symbol sequences encoding periodic orbits are periodic, and the symbol sequences of the collision orbits show a special symmetry. In Remark 6.12 we show how the *braid group* acts on the bounded orbits and how the bounded orbits can be seen in the scattering data.

The definition of *topological entropy* of a flow has been extended to the non-compact situation by Bowen, see Def. 7.1.

We want to estimate the topological entropy of the flow $\Phi_E^t = \Phi^t|_{\Sigma_E}$ on the energy shell. In Lemma 7.4 we show that the topological entropy is the same as the one of the motion restricted to the bounded orbits (this looks obvious but takes some time to prove).

Using this result and symbolic dynamics, we find in Thm. 7.6 bounds for the topological entropy which go like $\sqrt{E}$ if there are three or more nuclei. For n equal one or two the topological entropy vanishes.

The proliferation of the number $N_E(T)$ of closed orbits with minimal periods smaller than T is roughly exponential, with a rate given by the topological entropy. More detailed questions are discussed in *Chap. 8* . A statement for the number of fixed points of the Poincaré map is easily derived (8.1).

In Lemma 8.2 we find an interesting relation between the *desynchronization time* for geodesics and an integral over the symplectic two-form. This relation does not apply to more general Hamiltonian flows.

The number $\mathbf{N}_E(T)$ of closed geodesics grows asymptotically like

$$\mathbf{N}_E(T) \sim \frac{\exp(h_{\text{top}}T)}{h_{\text{top}}T}$$

(Prop. 8.3), and we derive a slightly weaker statement on the number $N_E(T)$ of closed orbits of the flow Φ_E^t in Theorem 8.5. There are related statements on zeta functions (8.4).

In *Chap. 9* we estimate for $n \geq 3$ nuclei the Hausdorff dimension of the set b_E of bounded orbits, because that quantity will turn out to be important in the context of a semiclassical analysis of quantum resonances [24]. In Thm. 9.4 we derive high energy bounds for $\dim_H(b_E)$ of the form $1 + C/\ln E$.

Time delay of a scattering orbit is defined in comparison with the motion in the Coulomb potential $-Z_\infty/|\vec{q}|$ in Def. 10.1 of *Chap. 10*. We estimate the energy-dependent *escape rate*; the measure $\kappa_E(T)$ of those orbits which have a time delay larger than T decays like

$$\exp\left(-C\sqrt{E}\ln E \cdot T\right),$$

as shown in Thm. 10.6.

Examples like the Yukawa potential show that the Liouville measure $\lambda(b_+)$ of the positive energy bound states $b_+ \in P$ is larger than zero in general. We discuss this question in *Chap. 11*.

The logarithm of the Coulomb potential is harmonic in two dimensions. Convex combinations of logarithmic harmonic potentials are logarithmic superharmonic, in the sense $\Delta \ln V \geq 0$, and for purely Coulombic potentials $\sum_{l=1}^{n} -Z_l/|\vec{q} - \vec{s}_l|$ with $n \geq 2$ the inequality is even strict (Remark 11.2).

If $\Delta \ln V \geq 0$, then the conclusions of Thm. 6.11 apply for all positive energies. In particular, the measure $\lambda(b_+)$ of the positive energy bounded orbits vanishes (Thm. 11.1).

In that case the motion for positive energy is integrable (this does not contradict the positivity of topological entropy, since we are in a non-compact situation).

For arbitrary Coulombic potentials we show that $\lambda(b_+) < \infty$, at least if the asymptotic charge $Z_\infty \neq 0$ (Thm. 11.3). For $Z_\infty = 0$ we must be more careful, since the measure of the low-energy bound states may diverge (Remark 11.4).

There is a relation between the measure $\lambda(b_+)$ of positive energy bound states, the positive part of the potential V, and the integrated time delay. This is a special case of the classical version of Levinson's theorem (Thm. 11.5). In particular, the integrated time delay vanishes for *all* ln-superharmonic potentials.

In *Chap. 12* we consider the differential cross section $\frac{d\sigma}{d\theta^+}(E, \theta^-, \theta^+)$ for energies E which give rise to a negative curvature $K_E \leq 0$ everywhere (the $\frac{d\sigma}{d\theta^+}$ depends on both the incoming and the outgoing angle $\theta^\pm$, since V is not radially symmetric in general).

In this situation we can classify the scattering orbits of given (θ^-, θ^+) using the semidirect product

$$\pi_1(\mathbf{M}) \star \mathbb{Z}_2.$$

In Thm. 12.1 we write the differential cross section in the form

$$\frac{d\sigma}{d\theta^+}(E,\theta^-,\theta^+) = -\frac{1}{\sqrt{2E}} \sum_{g\in(\pi_1(\mathrm{M})\star\mathbb{Z}_2)\setminus\{\mathrm{Id}\}} \left(\frac{d\varphi_{E,\theta^-}}{dL}(L_g^-(E,\theta^-,\theta^+))\right)^{-1}, \quad (1.3)$$

valid for $\theta^+ \neq \theta^-$.

The motion in a potential with n attracting coulombic singularities has been studied earlier by Bolotin [5] and Gutzwiller [14]. Bolotin proves nonintegrability (in the analytic sense) for $n > 2$ centres, whereas Gutzwiller discusses symbolic dynamics and semiclassical quantum mechanics.

In all examples of irregular scattering considered up to now, the differential cross section had an infinity of so-called *rainbow singularities* which were arranged in a fractal pattern (see Jung and Tél [20]). These rainbow singularities arise as densities at folds of the configuration space projection of the Lagrangean manifold composed of the orbits of a given incoming momentum.

Contrary to the previous examples, for $K_E \leq 0$ the differential cross section of Coulombic potentials is smooth, up to the forward direction (Thm. 12.3). For $Z_\infty > 0$, it has the same limiting behaviour for $\theta^+ \to \theta^-$ as the Rutherford cross section for a Coulomb potential of strength Z_∞ (Remark 12.4). In fact, the graphs of $\frac{d\sigma}{d\theta^+}$ for $n = 2$ and for the irregular cases $n \geq 3$ look very similar to the well-known Rutherford cross section for $n = 1$ (see Fig. 12.1).

So if nature were classical, the irregularity of the scattering process would not be seen in a differential cross section measurement. Somewhat ironically, Quantum Mechanics, which typically washes out effects of classical chaos, should lead to fluctuations in the differential cross section for $n \geq 3$, due to interference effects in the analog of (1.3).

Finally, in *Chap. 13*, we discuss possible future work concerning the model and its extensions.

We expect that many techniques developed in this work generalize to other systems exhibiting irregular scattering.

In particular, the use of the Jacobi metric allows one to take advantage of the many existing results on geodesic motion. The cone field technique seems to be a flexible tool for scattering problems.

Lemma 7.4, which relates the topological entropy for the energy shell to the more accessible entropy of the restriction of the flow to the bounded orbits, should immediately generalize to the typical applications in physics (but could be wrong for motion on negatively curved spaces, where the effect of the negative curvature at infinity may be dominant).

Notation. We have not chosen to be consistent in our notation, but there are some rules.

- Typically, references to energy appear as subscripts, whereas references to time appear as superscripts.
- The symbols of regions not containing (or projecting to) the locations $\vec{s}_l$ of the nuclei wear a hat. The same holds for functions defined on those regions.

- The branched covering surface $\mathbf{M}$ defined in (1.2) is basic to many constructions. In many cases there are pairs of objects of similar nature defined on the covering space and on the original space, respectively. Then the objects on the covering space have bold type symbols.

- The subindex ∞ often refers to objects connected with the Kepler motion which we use to define the Møller transformations.

- Symbols used within one proof are not defined outside that proof (this is especially important for constants).

- θ sometimes denotes the Heaviside function (which is one for positive values of its argument and zero otherwise); $[x]$ is the integer part of x.

- As usual, there is a problem concerning the meaning of *geodesic* (or geodesic segment) in this book. First of all, a geodesic is a map c from $\mathbb{R}$ to a Riemannian manifold M solving the geodesic equation. As such c is parametrized proportional to arc length, but not necessarily *by* arc length. Secondly, we sometimes call the image $c(\mathbb{R}) \subset M$ a geodesic, too.

  Furthermore we call the horizontal lift of a geodesic to the unit tangent bundle a *geodesic flow line*.

In the proofs we often change back and forth between the original configuration plane and the covering surface $\mathbf{M}$. That kind of reasoning may not seem to be very consistent, but to our opinion it is more effective.

Many statements are valid 'for large energies'. This means that there exists a threshold energy such that the statement holds true for all energies larger than that threshold.

Acknowledgement It was a great pleasure to learn from J. Moser about the work of Hadamard on Cantor sets generated by geodesic motion.

2. The Scattering Transformation

We consider the time evolution generated by a Hamiltonian

$$\hat{H}(\vec{q},\vec{p}) := \tfrac{1}{2}\vec{p}^{\,2} + V(\vec{q}) \tag{2.1}$$

with n negative Coulombic singularities of the potential $V(\vec{q})$ situated at the points $\vec{s}_1, \ldots, \vec{s}_n \in M := \mathbb{R}^2$ ($\vec{s}_i \neq \vec{s}_k$ for $i \neq k$). To control the asymptotic behaviour, we assume that V decomposes into the sum of a Coulombic potential and a short range potential. By this we mean the following:

Definition 2.1 A smooth, real-valued function V on the *configuration space*

$$\hat{M} := M \setminus \{\vec{s}_1, \ldots, \vec{s}_n\} \tag{2.2}$$

is called *negative Coulombic* if

1. There exist $Z_l > 0$, $l \in \{1, \ldots, n\}$, such that

$$V(\vec{q}) = \sum_{l=1}^{n} \frac{-Z_l}{|\vec{q} - \vec{s}_l|} + W(\vec{q}) \tag{2.3}$$

   with $W : M \to \mathbb{R}$ smooth. Z_l is called the *charge* of the l-th *nucleus.*

2. The potential vanishes at infinity, *i.e.*

$$\lim_{|\vec{q}| \to \infty} V(\vec{q}) = 0,$$

   and there exist $Z_\infty \in \mathbb{R}$, called the *asymptotic charge*, $\epsilon > 0$ and $R_{\min} > 1$ such that for all $\vec{q}, \vec{q}_1, \vec{q}_2 \in \hat{M}$ with $|\vec{q}|, |\vec{q}_1|, |\vec{q}_2| \geq R_{\min}$

$$\left| \nabla V(\vec{q}) - Z_\infty \frac{\vec{q}}{|\vec{q}|^3} \right| < |\vec{q}|^{-2-\epsilon} \tag{2.4}$$

   and

$$|\nabla V(\vec{q}_1) - \nabla V(\vec{q}_2)| < \frac{|\vec{q}_1 - \vec{q}_2|}{|\vec{q}_1|^{2+\epsilon}}. \tag{2.5}$$

Remarks 2.2

1. If the potential V meets condition 2.4, then the asymptotic charge Z_∞ is uniquely defined.

2. In the simplest example of a *purely Coulombic* potential

$$V(\vec{q}) := \sum_{l=1}^{n} \frac{-Z_l}{|\vec{q} - \vec{s}_l|}, \tag{2.6}$$

the asymptotic charge $Z_\infty = \sum_{l=1}^{n} Z_l$, but our definition also covers scattering by partially ionized, neutral ($Z_\infty = 0$) or negatively charged molecules.

3. For $n \geq 2$, a typical length scale in configuration space is given by

$$d_{\text{min}} := \min_{k \neq l} |\vec{s}_k - \vec{s}_l| \,. \tag{2.7}$$

In the case of a single atom ($n = 1$) we fix d_{min} by setting $d_{\text{min}} := 2R_{\text{min}}$, say.

One important threshold of the energy is

$$V_{\text{max}} := \sup_{\vec{q} \in \hat{M}} V(\vec{q}). \tag{2.8}$$

Clearly, $0 \leq V_{\text{max}} < \infty$, and in many cases $V_{\text{max}} = 0$.

Due to collisions with the nuclei situated at $\vec{s}_l$, the flow on the phase space $T^*\hat{M}$ of the particle is incomplete. There are several ways to regularize the collision orbits which are all essentially equivalent. In Chap. 3, we shall employ a regularization method which is particularily suited to understand the geometrical and topological problems arising, but which involves a time reparametrization of the phase space orbits.

In this chapter a time reparametrization is unwanted, since it would complicate the definition of the scattering transformation. Therefore, we use a different regularization. To preserve continuity of the motion with respect to the initial conditions, a particle colliding with a nucleus at $\vec{s}_l \in M$ must be reflected backwards. Therefore, we can parametrize the state of the colliding particle by its incoming (and outgoing) direction and by its energy. That is, we complete phase space by adjoining n cylinders $S^1 \times \mathbb{R}$, one for each nucleus.

For the case of a Coulomb potential we linearize the motion near collision by simply using as canonical coordinates angular momentum, the direction of the Runge-Lenz vector, energy and the time passed since the pericenter of the orbit. The first three of these functions are constant on the orbit. Then we add the cylinder of phase space points with angular momentum and time parameter both equal to zero.

Proposition 2.3 *There exists a unique smooth extension* (P,ω,H) *of the Hamiltonian system* $(T^*\hat{M}, d\vec{q} \wedge d\vec{p}, \hat{H})$*, where the* phase space P *is a smooth four-dimensional manifold with*

$$P := T^*\hat{M} \cup \bigcup_{l=1}^{n} \mathbb{R} \times S^1$$

as a set, ω *is a smooth symplectic two-form on* P *with*

$$\omega|_{T^*\hat{M}} = dq^1 \wedge dp_1 + dq^2 \wedge dp_2,$$

and $H : P \to \mathbb{R}$ *is a smooth Hamiltonian function with* $H|_{T^*\hat{M}} = \hat{H}$.

The smooth Hamiltonian flow

$$\Phi^t : P \to P, \quad t \in \mathbb{R} \tag{2.9}$$

generated by H *is complete.*

For all energies $E > V_{\max}$ *the* energy shell

$$\Sigma_E := \{x \in P \mid H(x) = E\} \tag{2.10}$$

is a smooth, three-dimensional manifold, and we write $\Phi_E^t := \Phi^t|_{\Sigma_E}$.

Proof. We introduce canonical coordinates to regularize the flow in the phase space neighbourhood $\hat{U}_l^\varepsilon$, $0 < \varepsilon < d_{\min}$, of the l-th nucleus, with

$$\hat{U}_l^\varepsilon := \left\{ (\vec{q},\vec{p}) \in T^*\hat{M} \,\middle|\, |\vec{q} - \vec{s}_l| < \varepsilon,\, \vec{p}^{\,2} > \frac{Z_l}{|\vec{q} - \vec{s}_l|} + 1 \right\}. \tag{2.11}$$

On $\hat{U}_l^\varepsilon$, the Hamiltonian function has the form $\hat{H}(\vec{q},\vec{p}) = \hat{H}_l(\vec{q},\vec{p}) + W_l(\vec{q})$ with

$$\hat{H}_l(\vec{q},\vec{p}) := \tfrac{1}{2}\vec{p}^{\,2} - \tfrac{Z_l}{|\vec{q}-\vec{s}_l|} \tag{2.12}$$

and the *smooth* additional potential W_l on $\{\vec{q} \in M \mid |\vec{q} - \vec{s}_l| < \varepsilon\}$,

$$W_l(\vec{q}) := \sum_{i \neq l} \frac{-Z_i}{|\vec{q} - \vec{s}_i|} + W(\vec{q}). \tag{2.13}$$

We first show how to regularize the purely Coulombic motion generated by $\hat{H}_l$ and then indicate the generalizations needed in the presence of the additional potential W_l.

Let $\hat{L}_l, \hat{T}_l : \hat{U}_l^\varepsilon \to \mathbb{R}$ and $\hat{\varphi}_l : \hat{U}_l^\varepsilon \to S^1$ be given by

$$\hat{L}_l(\vec{q},\vec{p}) := (\vec{q} - \vec{s}_l) \times \vec{p} = (q^1 - s_l^1)p_2 - (q^2 - s_l^2)p_1$$

$$\hat{T}_l(\vec{q},\vec{p}) := \int_{r_{\min}(\vec{q},\vec{p})}^{|\vec{q}-\vec{s}_l|} \frac{r}{\sqrt{2r^2\hat{H}_l(\vec{q},\vec{p}) + 2Z_l r - \hat{L}_l^2(\vec{q},\vec{p})}}\, dr \ \operatorname{sign}((\vec{q} - \vec{s}_l)\cdot\vec{p})$$

with

$$r_{\min}(\vec{q},\vec{p}) := \begin{cases} (-Z_l + \sqrt{Z_l^2 + 2\hat{H}_l(\vec{q},\vec{p})\hat{L}_l^2(\vec{q},\vec{p})})/2\hat{H}_l(\vec{q},\vec{p}) & , \hat{H}_l \neq 0 \\ \hat{L}_l^2(\vec{q},\vec{p})/2Z_l & , \hat{H}_l = 0 \end{cases} \tag{2.14}$$

and

$$\hat{\varphi}_l(\vec{q},\vec{p}) := \arg\left(-ip\hat{L}_l(\vec{q},\vec{p}) - Z_l \frac{q - s_l}{|\vec{q} - \vec{s}_l|}\right). \tag{2.15}$$

Here we denote by $v := v_1 + iv_2$ the complexification of a two-component vector $\vec{v}$.

$\hat{L}_l$ is the angular momentum relative to the position of the l-th nucleus.

$\hat{T}_l$ is the time elapsed since the closest encounter of the Kepler solution with the nucleus. There is only one such *pericenter* of the orbit, with distance $r_{\min}$, since $d((\vec{q} - \vec{s}_l) \cdot \vec{p})/dt > 0$ in $\hat{U}_l^\varepsilon$. $\hat{T}_l$ is a smooth function, which can be seen by explicit evaluation of the integrals (see Thirring [44], for more information):

$$\int \frac{r}{\sqrt{2r^2E + 2Zr - L^2}} dr = \tag{2.16}$$

$$\frac{r}{\sqrt{2E}}\sqrt{1 + \frac{Z}{rE} - \frac{L^2}{2r^2E}} - \frac{Z}{(2E)^{3/2}} \ln\left(Er + \tfrac{1}{2}Z + \sqrt{E(r^2E + Zr - \tfrac{1}{2}L^2)}\right)$$

for $E > 0$ and $Z > 0$.

$\hat{\varphi}_l$ coincides with the angle between the 1-direction and the direction of the pericenter. $\hat{\varphi}_l$ is a well-defined and smooth function, since the argument is non-zero:

$$\left| i\hat{L}_l p + Z_l \frac{q - s_l}{|\vec{q} - \vec{s}_l|} \right|^2 = 2\hat{L}_l^2 \hat{H}_l + Z_l^2 > 0.$$

This is obvious for positive energies, but for $\hat{H}_l < 0$ we have

$$2\hat{L}_l^2\hat{H}_l \geq \vec{p}^{\,2}|\vec{q} - \vec{s}_l|(\vec{p}^{\,2}|\vec{q} - \vec{s}_l| - 2Z_l) = (\vec{p}^{\,2}|\vec{q} - \vec{s}_l| - Z_l)^2 - Z_l^2 > -Z_l^2 \qquad \text{on } \hat{U}_l^\varepsilon.$$

The above coordinates are canonical: $\{\hat{T}_l, \hat{H}_l\} = \{\hat{\varphi}_l, \hat{L}_l\} = 1$, whereas all other Poisson brackets vanish. One easy way to see that $\{\hat{\varphi}_l, \hat{T}_l\} = 0$ is to observe that the Hamiltonian vector field $X_{\hat{\varphi}_l}$ of $\hat{\varphi}_l$ is tangential to the hypersurface $\hat{T}_l \equiv 0$. To this end we show that $\{\hat{\varphi}_l, (\vec{q} - \vec{s}_l) \cdot \vec{p}\} = 0$ on the surface $\hat{T}_l \equiv 0$ which equals

$$\hat{S}_l^\varepsilon := \{(\vec{q},\vec{p}) \in \hat{U}_l^\varepsilon \mid (\vec{q} - \vec{s}_l) \cdot \vec{p} = 0\}.$$

On this surface both complex numbers appearing in the r.h.s. of eq. (2.15) have the same argument (mod π) which is invariant under the dilation generated by $(\vec{q} - \vec{s}_l) \cdot \vec{p}$

On the other hand, $\{\hat{\varphi}_l, \hat{T}_l\}$ is invariant w.r.t. the flow ψ_l^t of $\hat{H}_l$:

$$\begin{aligned} \frac{d}{dt}(\psi_l^t)^*(\{\hat{\varphi}_l, \hat{T}_l\}) &= -(\psi_l^t)^*(\{\hat{H}_l, \{\hat{\varphi}_l, \hat{T}_l\}\}) \\ &= (\psi_l^t)^*(\{\hat{\varphi}_l, \{\hat{T}_l, \hat{H}_l\}\} + \{\hat{T}_l, \{\hat{H}_l, \hat{\varphi}_l\}\}) = 0, \end{aligned}$$

using the Jacobi identity. Up to collision orbits (which are of measure zero) every point in $\hat{U}_l^\varepsilon$ eventually meets $\hat{S}_l^\varepsilon$, which implies $\{\hat{\varphi}_l, \hat{T}_l\} = 0$ on $\hat{U}_l^\varepsilon$.

Thus by introducing the above coordinates, we obtain a canonical chart in $\hat{U}_l^\varepsilon$ which explicitly linearizes the $\hat{H}_l$ flow.

The motion is then regularized in the following way. One defines a completion of $\hat{U}_l^\varepsilon$ by setting $U_l^\varepsilon := \hat{U}_l^\varepsilon \cup \mathbb{R} \times S^1$ as a set, and one introduces a topology on U_l^ε by extending the coordinates $(\hat{H}_l, \hat{T}_l, \hat{L}_l, \hat{\varphi}_l)$ to coordinates $(H_l, T_l, L_l, \varphi_l) : U_l^\varepsilon \to \mathbb{R}^3 \times S^1$ by mapping a point $(h, \varphi) \in \mathbb{R} \times S^1$ in the cylinder onto $(h, 0, 0, \varphi)$. By that procedure we obtain the topological manifold P and, by taking limits, we extend the Hamiltonian $\hat{H}$ to a function $H : P \to \mathbb{R}$. We did not yet introduce a differential structure everywhere on P, since near the l-th cylinder that structure will depend on the germ of the additional potential W_l. Nevertheless the topology of P is already determined by the purely Coulombic local Hamiltonians $\hat{H}_l$, and, by taking limits for the collision orbits, we are able to extend the flow generated by $\hat{H}$ to a complete continuous flow Φ^t on P (it is clear that the particle cannot escape to spatial infinity in finite time).

To generalize the construction to the case of the flow generated by H which is of the local form $H_l + W_l$, we define similar canonical coordinates in U_l^ε which linearize the H-flow Φ^t.

We start by observing that for $\varepsilon > 0$ small enough, Φ^t is transversal to the hypersurface $\hat{S}_l^\varepsilon$ since

$$\begin{aligned} \frac{d}{dt}(\vec{q} - \vec{s}_l)\cdot\vec{p} &= \vec{p}^{\,2} - (\vec{q} - \vec{s}_l) \cdot \nabla V(\vec{q}) \\ &= \vec{p}^{\,2} - \frac{Z_l}{|\vec{q} - \vec{s}_l|} - (\vec{q} - \vec{s}_l) \cdot \nabla W_l(\vec{q}) > 1 - \varepsilon\,|\nabla W_l(\vec{q})| > \tfrac{1}{2} \end{aligned}$$

by def. (2.11) of $\hat{U}_l^\varepsilon$ and smoothness of the potential W_l.

We extend that hypersurface to the topological submanifold $S_l^\varepsilon := \hat{S}_l^\varepsilon \cup \mathbb{R} \times S^1 \subset U_l^\varepsilon$. We define functions

$$(\tilde{H}_l, \tilde{T}_l, \tilde{L}_l, \tilde{\varphi}_l) : U_l^\varepsilon \to \mathbb{R}^3 \times S^1$$

by letting $\tilde{H}_l := H|_{U_l^\varepsilon}$ be the energy and

$$\tilde{T}_l(\Phi^t(x)) := t, \quad x \in S_l^\varepsilon \tag{2.17}$$

the time passed since the passage of the pericenter. Note that (2.17) defines $\tilde{T}_l$ everywhere on U_l^ε, since every orbit in U_l^ε passes S_l^ε exactly once.

$\tilde{L}_l(\Phi^t(x)) := L_l(x)$ and $\tilde{\varphi}_l(\Phi^t(x)) := \varphi_l(x)$, $x \in S_l^\varepsilon$, are then the angular momentum and the Runge-Lenz angle at the pericenter x of the orbit.

Clearly, *by fiat*, $\tilde{H}_l, \tilde{L}_l$ and $\tilde{\varphi}_l$ are constant on one orbit $\Phi^t(\vec{q}, \vec{p})$, whereas $\tilde{T}_l(\Phi^t(\vec{q}, \vec{p}))$ is an affine function of time t. Therefore we have linearized the full motion. We must show that the functions $(\tilde{H}_l, \tilde{T}_l, \tilde{L}_l, \tilde{\varphi}_l)$, restricted to $\hat{U}_l^\varepsilon$, are indeed smooth coordinates. One easy way to do that is to use the Levi-Civita covering construction. Therefore, we postpone that smoothness proof to Proposition 3.1.

We have $\{\tilde{T}_l, \tilde{H}_l\} = \{\tilde{\varphi}_l, \tilde{L}_l\} = 1$ with the other Poisson brackets vanishing in $\hat{U}_l^\varepsilon$, as follows from the Hamiltonian flow box theorem (Thm. 5.2.19 of [1]).

Since the coordinates $(\tilde{H}_l, \tilde{T}_l, \tilde{L}_l, \tilde{\varphi}_l)$ define a homeomorphism of U_l^ε onto its image, we use them to define a differential structure on the whole of U_l^ε and thus on P. The symplectic two-form ω is then defined on U_l^ε by setting

$$\omega|_{U_l^\varepsilon} := d\tilde{T}_l \wedge d\tilde{H}_l + d\tilde{\varphi}_l \wedge d\tilde{L}_l.$$

The symplectic manifold (P, ω) is constructed by performing that regularization near every nucleus.

Smoothness of the energy shells Σ_E for $E > V_{\max}$ follows by noticing that these E are regular values of $H : P \to \mathbb{R}$.

Notation 2.4 We will denote by $\eta : P \to M$ the projection to the configuration plane, *i.e.* $\eta(\vec{q}, \vec{p}) := \vec{q}$ for $(\vec{q}, \vec{p}) \in T^*\hat{M}$ and $\eta(x) := \vec{s}_l$ if x lies in the l-th cylinder.

As this notation is somewhat heavy, we will use the standard notation $(\vec{q}, \vec{p})$ for an arbitrary point in phase space P whenever this cannot lead to confusion.

Furthermore, although neither $T^*\hat{M}$ nor P is a vector space, we can, with the above notation, use the canonical injection

$$T^*\hat{M} \ni x \mapsto (\vec{q}, \vec{p}) \in \mathbb{R}^4$$

to define a metric on $T^*\hat{M}$ by setting

$$\mathrm{dist}(x_1, x_2) = |(\vec{q}_1, \vec{p}_1) - (\vec{q}_2, \vec{p}_2)|, \tag{2.18}$$

where $|\cdot|$ denotes the Euclidean distance in $\mathbb{R}^4$.

On $T^*\hat{M}$ the metric topology of $\mathrm{dist}(\cdot,\cdot)$ coincides with the topology of P, and we shall henceforth freely use the rhs. of (2.18) to denote this distance function on the $T^*\hat{M}$-part of P.

Now we are ready to introduce the classical analogues of the quantum mechanical Møller operators and the scattering matrix, following the work of Simon [39] and Hunziker [19].

Since the force $-\nabla V(\vec{q})$ goes to zero for large $|\vec{q}|$, one expects the asymptotic motion to be comparable with the free motion generated by the Hamiltonian function $\vec{p}^2/2$. As long as one is only interested in the scattering direction depending on the initial data, this is indeed the case. But here we want to study *all* asymptotic observables, in particular the time delay, which is important for the understanding of the corresponding quantum system in the semiclassical limit. As is well-known, the time delay diverges logarithmically for the motion in a Coulomb potential generated by

$$\hat{H}_\infty(\vec{q}, \vec{p}) := \tfrac{1}{2}\vec{p}^2 - \tfrac{Z_\infty}{|\vec{q}|} \tag{2.19}$$

if $Z_\infty \neq 0$. Up to short-range corrections, our potential equals $-Z_\infty / |\vec{q}|$, which makes it impossible to define the Møller transformations by comparing with free motion.

Instead, we use the smooth complete flow

$$\Phi^t_\infty : P_\infty \to P_\infty \tag{2.20}$$

generated by (2.19). If $Z_\infty = 0$, Φ^t_∞ is the free flow on the phase space $P_\infty := T^*\mathbb{R}^2$, but if $Z_\infty > 0$, we have to regularize $T^*(\mathbb{R}^2 \setminus \{0\})$ in the way described in Proposition 2.3 to obtain P_∞. Finally, if $Z_\infty < 0$, then Φ^t_∞ is already complete on $P_\infty := T^*(\mathbb{R}^2 \setminus \{0\})$, since particles of finite energy cannot meet the origin at $\vec{q} = 0$.

Thus we are to compare motions Φ^t and Φ^t_∞ on the *different* phase spaces P and P_∞. We cannot just identify P with P_∞ by neglecting the measure zero sets projecting to the singularities, since certain sets of measure zero will turn out to be crucial in our analysis of the typical behaviour.

We overcome the above difficulty by observing that it suffices to identify P_∞ with P in a neighbourhood of spatial infinity. More precisely, let

$$P_{\infty,+} := \{x \in P_\infty \mid H_\infty(x) > 0\} \tag{2.21}$$

be the set of phase space points with positive "free" energy. Then the orbit $\Phi^t_\infty(x)$ starting at $x \in P_{\infty,+}$ goes to spatial infinity for large positive and negative times. The ball $\{\vec{q} \in M \mid |\vec{q}| \leq R_{\min}\}$ contains all singularities of V and the singularity at the origin of the Coulomb Hamiltonian H_∞. Therefore, we can canonically identify points $(\vec{q}, \vec{p}) \in P_\infty$ with points $(\vec{q}, \vec{p}) \in P$ if $|\vec{q}| > R_{\min}$, and we denote this identification by Id. Therefore, we conclude that the *Møller transformation*

$$\Omega^\pm := \lim_{t \to \pm\infty} \Phi^{-t} \circ \mathrm{Id} \circ \Phi^t_\infty \tag{2.22}$$

are formally maps $\Omega^\pm : P_{\infty,+} \to P$. Now we are going to prove the existence of the pointwise limit in (2.22).

First some standard definitions (see [19]):

Definition 2.5

$$\begin{aligned}
b^{\pm,k} &:= \left\{x \in P \mid \left|\eta(\Phi^t(x))\right| \leq k \text{ for } 0 \leq \pm t < \infty \text{ and } |H(x)| \leq k\right\} \\
b^\pm &:= \bigcup_{k \in \mathbb{N}} b^{\pm,k} \\
b &:= b^+ \cap b^- \qquad \text{(the \textit{bound states})} \\
s^\pm &:= \{x \in P \mid x \notin b^\pm \text{ and } H(x) > 0\} \\
s &:= s^+ \cap s^- \qquad \text{(the \textit{scattering states}).}
\end{aligned}$$

We shall show that $s^\pm = \Omega^\pm(P_{\infty,+})$ so that the term 'scattering states' is really justified.

By continuity of Φ^t, the $b^{\pm,k}$ are compact, hence $b^\pm$ and $s^\pm$ are measurable and Φ^t-invariant.

Before proving the existence of the Møller transformations, we show the existence of their inverse. This is a somewhat weaker statement since a Lipschitz condition of the form (2.5) is not needed.

In this book we shall repeatedly apply one important consequence of the *virial identity*

$$\frac{d}{dt}(\vec{q}(t)\cdot\vec{p}(t)) = 2(E - V(\vec{q}(t))) - \vec{q}(t)\cdot\nabla V(\vec{q}(t)) \tag{2.23}$$

which holds true for any orbit $(\vec{q}(t),\vec{p}(t)) = \Phi^t(\vec{q}_0,\vec{p}_0)$ with energy $E := H(\vec{q}_0,\vec{p}_0)$ (whenever $\vec{q}(t) \neq \vec{s}_l$). Let us choose a function $R_{\mathrm{vir}} : (0,\infty) \to \mathbb{R}$ of the energy, called the *virial radius*, with

$$|V(\vec{q})| < E/2 \text{ and } |\vec{q}\cdot\nabla V(\vec{q})| < E/2 \text{ for all } \vec{q} \text{ with } |\vec{q}| \geq R_{\mathrm{vir}}(E). \tag{2.24}$$

As a consequence of part 2 of Def. 2.1 of Coulombic potentials, such a function $R_{\mathrm{vir}} \geq R_{\mathrm{min}}$ exists. W.l.o.g. we assume R_{vir} to be smooth, nonincreasing and constant for large energies. Then by (2.23) and (2.24)

$$\frac{d}{dt}(\vec{q}(t)\cdot\vec{p}(t)) > \frac{E}{2} > 0 \quad \text{if } E := H(\vec{q}_0,\vec{p}_0) > V_{\mathrm{max}} \text{ and } |\vec{q}(t)| \geq R_{\mathrm{vir}}(E). \tag{2.25}$$

In particular a trajectory $\vec{q}(t)$ leaving a ball of radius $R_{\mathrm{vir}}(E)$ cannot reenter this ball in the future but must go to spatial infinity.

Proposition 2.6 *The limits*

$$\lim_{t\to\pm\infty} \Phi_\infty^{-t} \circ \mathrm{Id}^{-1} \circ \Phi^t \tag{2.26}$$

exist pointwisely on $s^\pm \subset P$ and thus define maps $\Omega_*^\pm : s^\pm \to P_{\infty,+}$ intertwining Φ^t and Φ_∞^t:

$$\Omega_*^\pm \circ \Phi^t = \Phi_\infty^t \circ \Omega_*^\pm. \tag{2.27}$$

In particular, the asymptotic limits $\vec{p}^\pm : s^\pm \to \mathbb{R}^2$ and $L^\pm : s^\pm \to \mathbb{R}$ of the momentum and the angular momentum $L(\vec{q},\vec{p}) := \vec{q}\times\vec{p}$

$$\vec{p}^\pm(\vec{q}_0,\vec{p}_0) := \lim_{t\to\pm\infty} \vec{p}(t) \text{ and } L^\pm(\vec{q}_0,\vec{p}_0) := \lim_{t\to\pm\infty} L(\vec{q}(t),\vec{p}(t)) \tag{2.28}$$

exist, for $(\vec{q}_0,\vec{p}_0) \in s^\pm$ and $(\vec{q}(t),\vec{p}(t)) = \Phi^t(\vec{q}_0,\vec{p}_0)$.

Proof. We closely follow the lines of [19] and [39]. First we show (2.28) which fixes the trajectory and allows to prove the full statement (2.26). We consider only the case $t \to +\infty$.

Now let $(\vec{q}_0,\vec{p}_0) \subset s^+$ and $E := H(\vec{q}_0,\vec{p}_0)$. Then there exists a time t_0 such that $|\vec{q}(t_0)| > R_{\mathrm{vir}}(E)$. Furthermore, we can assume that $\vec{q}(t_0)\cdot\vec{p}(t_0) \geq 0$, since otherwise $|\vec{q}(t)|$ would be uniformly bounded for $t \geq t_0$.

But then $\vec{q}(t)\cdot\vec{p}(t) \geq (t-t_0)E/2$ for $t \geq t_0$, as follows from (2.24) and (2.23). This implies

$$|\vec{q}(t)| \geq \max\left(R_{\mathrm{vir}}(E), (t-t_0)\sqrt{E}/4\right) \tag{2.29}$$

for $t \geq t_0$. Thus by (2.4)

$$\left|\frac{d}{dt}\vec{p}(t)\right| = |\nabla V(\vec{q}(t))| < \frac{16(1+|Z_\infty|)}{E}(t-t_0)^{-2} \qquad \text{for } t \geq t_0$$

and the limit $\vec{p}_+(\vec{q}_0, \vec{p}_0) = \lim_{t\to\infty} \vec{p}(t)$ exists.

Next we show that the asymptotic limit $L^+(\vec{q}_0, \vec{p}_0)$ of the angular momentum exists. For $t \geq t_0$ we can estimate

$$\begin{aligned} \left|\frac{d}{dt}L(\vec{q}(t), \vec{p}(t))\right| &= |\nabla V(\vec{q}(t)) \times \vec{q}(t)| \\ &= \left|\left(\nabla V(\vec{q}(t)) - Z_\infty \frac{\vec{q}}{|\vec{q}|^3}\right) \times \vec{q}(t)\right| \\ &\leq |\vec{q}(t)|^{-1-\epsilon} \leq \left|(t-t_0)\sqrt{E}/4\right|^{-1-\epsilon} \end{aligned}$$

using (2.4) and (2.29), which shows the existence of the limit $L^+(\vec{q}_0, \vec{p}_0)$.

The only thing left is to control the distance between $\vec{q}(t)$ and a point moving on a Kepler solution which is orbitally asymptotic to $\vec{q}(t)$, *i.e.* $(\vec{Q}(t), \vec{P}(t)) := \Phi^t_\infty(\vec{Q}_0, \vec{P}_0)$ with

$$\lim_{t\to\infty} \vec{P}(t) = \vec{p}^{\,+}(\vec{q}_0, \vec{p}_0) \text{ and } L(\vec{Q}_0, \vec{P}_0) = L^+(\vec{q}_0, \vec{p}_0). \tag{2.30}$$

The radial coordinates $r(t) := |\vec{q}(t)|$ and $R(t) := \left|\vec{Q}(t)\right|$ meet the equations

$$\dot{r}^2(t) = 2(E - V(\vec{q}(t)) - (\vec{q}(t) \times \vec{p}(t)/r(t))^2$$

and

$$\dot{R}^2(t) = 2(E + Z_\infty/R(t)) - (L^+/R(t))^2.$$

Therefore, their difference

$$\dot{r}^2(t) - \dot{R}^2(t) = -2(Z_\infty/R(t) + V(\vec{q}(t))) + (L^+/R(t))^2 - (\vec{q}(t) \times \vec{p}(t)/r(t))^2$$

can be estimated for t large by

$$\begin{aligned} |\dot{r}^2(t) - \dot{R}^2(t)| &\leq 2|Z_\infty||1/r(t) - 1/R(t)| + \mathcal{O}(t^{-1-\epsilon}) \\ &\leq ct^{-2}|r(t) - R(t)| + \mathcal{O}(t^{-1-\epsilon}), \end{aligned} \tag{2.31}$$

using the estimate (2.29) for $r(t)$, and a similar estimate for $R(t)$, (2.30) and (2.4).

Dividing (2.31) by $|\dot{r}(t) + \dot{R}(t)| \geq \sqrt{E}$, and denoting $r(t) - R(t)$ by $w(t)$, we obtain the estimate

$$|\dot{w}(t)| \leq cE^{-1/2}t^{-2}|w(t)| + \mathcal{O}(t^{-1-\epsilon})$$

from which we conclude that $w^+ := \lim_{t\to\infty} w(t)$ exists. Therefore,

$$\lim_{t\to\infty} \left|\vec{q}(t) - \vec{Q}(t + w^+/\sqrt{2E})\right| = 0.$$

Thus we have shown the existence of a phase space point $(\vec{Q}_0, \vec{P}_0)$ with

$$\lim_{t\to\infty} \left|\Phi^t_\infty(\vec{Q}_0, \vec{P}_0) - \Phi^t(\vec{q}_0, \vec{p}_0)\right| = 0.$$

Furthermore, $\lim_{t\to\infty} \left|\Phi^t_\infty(\vec{Q}_1, \vec{P}_1) - \Phi^t_\infty(\vec{Q}_0, \vec{P}_0)\right| = 0$ only for $(\vec{Q}_1, \vec{P}_1) = (\vec{Q}_0, \vec{P}_0)$ which completes the proof. □

Now we show existence of the Møller transformations (2.22) following [39] and [19], with $\Omega^\pm = (\Omega^\pm_*)^{-1}$.

In [39], Simon showed in the context of short-range potentials that a Lipschitz condition of type (2.5) for the asymptotic behaviour of the forces is needed to ensure that the inverse Møller transformations $\Omega^\pm_*$ are one-to-one.

Proposition 2.7 *The limits*

$$\Omega^\pm := \lim_{t\to\pm\infty} \Phi^{-t} \circ \mathrm{Id} \circ \Phi^t_\infty$$

exist pointwisely on $P_{\infty,+} \subset P_\infty$ and thus define the Møller transformations $\Omega^\pm : P_{\infty,+} \to s^\pm$. These are continuous, measure-preserving, and intertwine Φ^t and Φ^t_∞:

$$\Omega^\pm \circ \Phi^t_\infty = \Phi^t \circ \Omega^\pm.$$

Proof. We show the assertions for Ω^+. For $x \in P_{\infty,+}$ there exists a time t_0 such that $\left|\vec{Q}(t)\right| > 2R_{\mathrm{vir}}(E)$ for all $t > t_0$, with $(\vec{Q}(t), \vec{P}(t)) := \Phi^t_\infty(x)$, and $E := H_\infty(x)$. In that region, far away from the nuclei, the difference between V and $Z_\infty/\left|\vec{q}\right|$ is small. Therefore, we write a solution curve $\vec{q}(t)$ of $\ddot{\vec{q}}(t) = -\nabla V(\vec{q}(t))$ in the form

$$\vec{q}(t) = \vec{Q}(t) + \vec{r}(t).$$

$\vec{r}(t)$ then is a solution of the differential equation

$$\ddot{\vec{r}}(t) = -\nabla V(\vec{Q}(t) + \vec{r}(t)) + Z_\infty \frac{\vec{Q}(t)}{\left|\vec{Q}(t)\right|^3},$$

which we transform into an integral equation. For a large enough minimal time $t_1 \geq t_0$, we know that $\left|\vec{Q}(t)\right| > 2R_{\mathrm{vir}}(E) + \sqrt{E}\cdot(t - t_1)/4$ for all $t \geq t_1$, see (2.29). Let

$$\mathcal{C}_T := \left\{\vec{r} \in C([T,\infty), \mathbb{R}^2) \,\middle|\, \sup_t |\vec{r}(t)| < 1\right\}.$$

Then

$$(\mathcal{F}\vec{r})(t) := \int_t^\infty ds \int_s^\infty d\tau \left(-\nabla V(\vec{Q}(\tau) + \vec{r}(\tau)) + Z_\infty \frac{\vec{Q}(\tau)}{\left|\vec{Q}(\tau)\right|^3} \right) \tag{2.32}$$

is well-defined for $\vec{r} \in \mathcal{C}_{t_1}$, since the integrand is bounded in absolute value by $\left|\vec{Q}(\tau)\right|^{-2-\epsilon} < 2(2R_{\text{vir}}(E) + \sqrt{E} \cdot (\tau - t_1)/4)^{-2-\epsilon}$, using (2.4) and (2.5). Therefore for a suitable $T \geq t_1$, $\mathcal{F}$ can be considered as a map $\mathcal{F} : \mathcal{C}_T \to \mathcal{C}_T$, which furthermore is a contraction, since

$$|(\mathcal{F}\vec{r}_1)(t) - (\mathcal{F}\vec{r}_2)(t)| \leq \int_t^\infty ds \int_s^\infty d\tau \left|\vec{Q}(\tau) + \vec{r}_1(\tau)\right|^{-2-\epsilon} |\vec{r}_1(\tau) - \vec{r}_2(\tau)| \tag{2.33}$$

for $t \geq t_1$, using (2.5). Therefore, $\mathcal{F}$ has a unique fixed point $\vec{r} \in \mathcal{C}_T$. The rest of the proof proceeds in complete analogy with [39]. □

Corollary 2.8

1. *$s^\pm = \Omega^\pm(P_{\infty,+})$, that is, every positive energy orbit which is unbounded in positive (negative) time is positively (negatively) asymptotic to a Kepler hyperbola.*

2. *The motion Φ^t generated by the Hamiltonian function $H : P \to \mathbb{R}$ is* asymptotically complete, *that is, up to a subset of Liouville measure zero the phase space consists of bound states and scattering states:*

$$\lambda(P \setminus b \cup s) = 0.$$

   Here we define Liouville measure by $\lambda := \frac{1}{2}\omega \wedge \omega$.

3. *The* scattering transformation

$$S := \Omega_*^+ \circ \Omega^- : D \to P_\infty \tag{2.34}$$

   with domain $D := \Omega_*^-(s) \subset P_{\infty,+}$ and range $\Omega_*^+(s)$ is continuous and Φ_∞^t-invariant, i.e.

$$S \circ \Phi_\infty^t = \Phi_\infty^t \circ S.$$

Proof.

1. The equality of $s^\pm$ with $\Omega^\pm(P_{\infty,+})$ is a consequence of Proposition 2.6, since $\Omega_*^\pm = (\Omega^\pm)^{-1}$ is defined on $s^\pm$.

2. This is a consequence the fact that $\Phi^t(b^{+,k}) \subset b^{+,k}$ for $t \geq 0$ but $\lambda(\Phi^t(b^{+,k})) = \lambda(b^{+,k})$, since Φ^t is canonical (see [19]).

3. Follows from Propositions 2.6 and 2.7. □

Proposition 2.9 *Let V be a Coulombic potential whose partial derivatives decay at infinity according to*

$$\lim_{\vec{q}\to\infty} |\vec{q}|^{|n|+1+\epsilon} \left| \frac{\partial^n}{\partial q^n} \left(V(\vec{q}) + \frac{Z_\infty}{|\vec{q}|} \right) \right| = 0 \qquad \forall n \in \mathbb{N}_0 \times \mathbb{N}_0. \tag{2.35}$$

Then the Møller transformations $\Omega^\pm$ are smooth canonical transformations.

Proof. It suffices to show that the contraction $\mathcal{F}$ depends smoothly on x, since in such a situation, the unique fixed point of $\mathcal{F}$ depends smoothly on the parameter x, too (see, e.g., Loomis and Sternberg [31], Chap. 4.9).

If the set of conditions (2.35) is met by the Coulombic potential V, then the partial derivatives of the integrand in (2.33) w.r.t. the initial conditions $x \in P_{\infty,+}$ are majorized by a multiple of $\tau^{-2-\epsilon}$ in absolute value. This is a result of the fact that nearby orbits of the Coulomb motion diverge in time only linearly. □

Remarks 2.10

1. Clearly, under these conditions, the Møller and scattering transformations are diffeomorphisms onto their images, as follows from the implicit mapping theorem.

2. Condition (2.35) is met by the purely Coulombic potentials,

$$V(\vec{q}) = \sum_{l=1}^{n} \frac{-Z_l}{|\vec{q} - \vec{s}_l|}.$$

3. Physical considerations show that (2.35) is not as restrictive as it may look like. One may think of V of being the restriction to a plane of the potential of a quantum mechanical molecule whose nuclei lie in that plane.

   As shown by Agmon and others, the wave functions of the electrons bound by the nuclei decay exponentially with the distance together with their partial derivatives. Therefore, the large-distance form of the potential induced by the nuclei and the charge density of the bound electrons should meet the assumptions of Proposition 2.9.

3. Regularization

In Chap. 2 we regularized the motion near the nuclei using a canonical transformation linearizing that motion. Compared to that method, the regularization scheme employed in this chapter has several disadvantages: It leads to a reparametrization of time; it works only for a fixed value of the energy; and the configuration plane M is replaced by a surface $\mathbf{M}$ covering M with branch points.

Yet this regularization turns out to be useful for one important reason: It allows to analyse the motion in phase space using Riemannian geometry which in this case is intimately linked to the topological aspects of the motion.

Our first step is to define the so-called *Jacobi metric* $\hat{g}_E$ on configuration space $\hat{M}$ (see, e.g., Abraham and Marsden [1], Chap. 3.7), which is conformally equivalent to the Euclidean metric $\hat{g}$ on $\hat{M}$ inherited from $M = \mathbb{R}^2$ (see Def. (2.2)):

$$\hat{g}_E(\vec{q}) := (1 - V(\vec{q})/E)\hat{g}(\vec{q}), \qquad \vec{q} \in \hat{M}, \tag{3.1}$$

where $E > V_{\mathbf{max}}$ is the energy of the particle whose motion we want to analyse. Up to a reparametrization of time, the solution curves of the geodesic motion on $(\hat{M}, \hat{g}_E)$ coincide with the motion $\eta_E \circ \Phi_E^t$ generated by the Hamiltonian H for initial conditions with energy E. Here $\eta_E := \eta|_{\Sigma_E}$ denotes the projection to the configuration plane.

On the other hand, $(\hat{M}, \hat{g}_E)$ is geodesically incomplete. Therefore we define the branched covering surface $\mathbf{M}$ of $M \cong \mathbb{C}$ as the concrete Riemann surface

$$\mathbf{M} := \left\{ (q, Q) \in \mathbb{C} \times \mathbb{C} \,\middle|\, Q^2 = \prod_{l=1}^{n} (q - s_l) \right\}, \tag{3.2}$$

using the identification of vectors $\vec{q} \in M$ with points $q \in \mathbb{C}$. Later on we shall forget the embedding of $\mathbf{M}$ in $\mathbb{C}^2$ and consider $\mathbf{M}$ as an abstract complex manifold with generic points denoted by $q \in \mathbf{M}$. Then the projection

$$\pi : \mathbf{M} \to M, \quad (q, Q) \mapsto q,$$

is a two-sheeted branched covering whose *branch points* $\mathbf{s}_l := (s_l, 0)$, $l \in \{1, \ldots, n\}$, are of order one and project to the positions s_l of the nuclei.

The identity map and

$$G : \mathbf{M} \to \mathbf{M}, \qquad (q, Q) \mapsto (q, -Q) \tag{3.3}$$

are the only *covering transformations*, that is, continuous transformations of $\mathbf{M}$ leaving the projection invariant: $\pi \circ G = \pi$.

Denoting by $\hat{\mathbf{M}}$ the Riemann surface

$$\hat{\mathbf{M}} := \mathbf{M} \setminus \{\text{branch points}\},$$

the restriction of π to $\hat{\mathbf{M}}$ leads to a two-fold *unbranched* covering $\hat{\pi} : \hat{\mathbf{M}} \to \hat{M}$. Thus we may lift the Jacobi metric $\hat{g}_E$ to the metric $\hat{\mathbf{g}}_E := \hat{\pi}^* \hat{g}_E$ on $\hat{\mathbf{M}}$.

The form of geodesics does not depend on their (constant) speed. Thus we assume their speed to be one. Then we obtain a geodesic flow on the unit tangent bundle $\hat{\boldsymbol{\Sigma}}_E := \{X \in T\hat{\mathbf{M}} \mid \hat{\mathbf{g}}_E(X,X) = 1\}$ (which is still incomplete).

To relate the geodesic flow on $\hat{\boldsymbol{\Sigma}}_E$ to the restriction of the flow Φ^t to the energy shell $\hat{\Sigma}_E := \{(\vec{q},\vec{p}) \in T^*\hat{M} \mid \hat{H}(\vec{q},\vec{p}) = E\}$, we use a two-sheeted smooth unbranched covering

$$\hat{\pi}_E : \hat{\boldsymbol{\Sigma}}_E \to \hat{\Sigma}_E, \qquad (q,\dot{q}) \mapsto \left(\pi(q), \sqrt{2E}(1 - V(\pi(q))/E) \cdot T_q\pi(\dot{q})\right), \tag{3.4}$$

that is, we project the point $(q,\dot{q}) \in \hat{\boldsymbol{\Sigma}}_E$ to the tangent space of the configuration space $\hat{M}$, rescale the speed (and implicitly identify velocity with momentum, using the Euclidean metric $\hat{g}$).

Proposition 3.1 *For $E > V_{\max}$ there exists a unique extension of the metric $\hat{\mathbf{g}}_E$ on $\hat{\mathbf{M}}$ to a smooth metric $\mathbf{g}_E$ on $\mathbf{M}$. The covering transformation $G : \mathbf{M} \to \mathbf{M}$ lifts to a fixed-point free diffeomorphism $G^* : \boldsymbol{\Sigma}_E \to \boldsymbol{\Sigma}_E$ on the unit tangent bundle*

$$\boldsymbol{\Sigma}_E := \{X \in T\mathbf{M} \mid \mathbf{g}_E(X,X) = 1\}, \qquad \boldsymbol{\eta}_E : \boldsymbol{\Sigma}_E \to \mathbf{M} \tag{3.5}$$

of the complete Riemann surface $(\mathbf{M}, \mathbf{g}_E)$. The geodesic flow

$$\boldsymbol{\Phi}_E^t : \boldsymbol{\Sigma}_E \to \boldsymbol{\Sigma}_E, \qquad t \in \mathbb{R} \tag{3.6}$$

on $\boldsymbol{\Sigma}_E$ is G^*-invariant: $\boldsymbol{\Phi}_E^t \circ G^* = G^* \circ \boldsymbol{\Phi}_E^t$ for all $t \in \mathbb{R}$.

There exists a unique two-fold smooth unbranched covering $\pi_E : \boldsymbol{\Sigma}_E \to \Sigma_E$ of the energy shell $\Sigma_E = \{x \in P \mid H(x) = E\}$ such that $\pi_E|_{\hat{\boldsymbol{\Sigma}}_E} = \hat{\pi}_E$. π_E is G^*-invariant, i.e. *$\pi_E \circ G^* = \pi_E$, so that it defines a homeomorphism $\boldsymbol{\Sigma}_E/\mathbb{Z}_2 \to \Sigma_E$. The diagram*

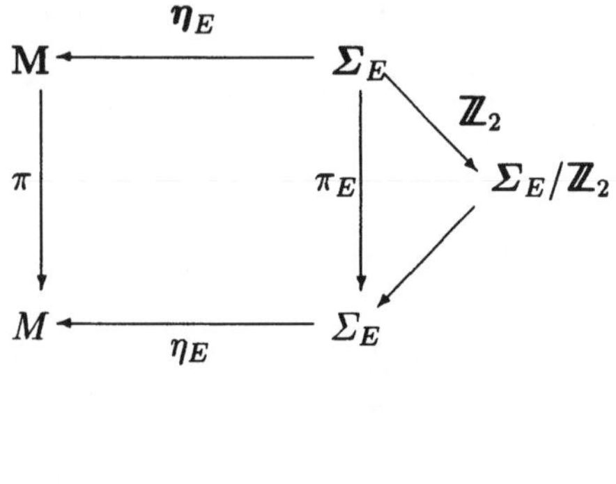

commutes.

The time reparametrization $s : \mathbb{R} \times \Sigma_E \to \mathbb{R}$ *defined by*

$$s(t,x) := \frac{1}{\sqrt{2E}} \int_0^t \frac{E\,dt'}{E - V(\eta_E \circ \pi_E \circ \Phi_E^{t'}(x))} \tag{3.7}$$

is smooth and

$$\forall t \in \mathbb{R}\, \forall x \in \Sigma_E : \pi_E \circ \Phi_E^t(x) = \Phi_E^{s(t,x)} \circ \pi_E(x). \tag{3.8}$$

Proof. By performing an analytic change of charts in M and $\mathbf{M}$ in the vicinity of the l-th branch point $\mathbf{s}_l$, we may assume w.l.o.g. that locally (3.2) attains *exactly* the form $Q^2 = q$ of the *Levi-Civita transformation* (see Farkas and Kra [11, I.1.6] for the simple construction of such a chart).

We consider the explicit form of the lifted metric $\hat{\mathbf{g}}_E = \hat{\pi}^* \hat{g}_E$. With

$$\hat{g}_E(q) = (1 - V(q)/E)(dq_1 \otimes dq_1 + dq_2 \otimes dq_2)$$

we have

$$\begin{aligned} \hat{\mathbf{g}}_E(Q) &= 4Q\bar{Q}(1 - V(Q^2)/E)(dQ_1 \otimes dQ_1 + dQ_2 \otimes dQ_2) \\ &= 4\left(Z_l/E + Q\bar{Q}(1 - W_l(Q^2)/E)\right)(dQ_1 \otimes dQ_1 + dQ_2 \otimes dQ_2) \end{aligned}$$

with W_l defined in (2.13). Obviously we can extend this equation to the branch point ($Q = 0$). Applying this prescription at each branch point, we end up with a smooth metric $\mathbf{g}_E$ on $\mathbf{M}$. $\mathbf{g}_E$ is complete since the conformal factor $(1 - V(q)/E)$ in the definition of the Jacobi metric is strictly positive and asymptotically constant for $E > V_{\max}$ and $|\vec{q}| \to \infty$.

It is clear that G is an isometry on $(\mathbf{M}, \mathbf{g}_E)$, since

$$G^* \hat{\mathbf{g}}_E = G^*(\hat{\pi}^* \hat{g}_E) = (\hat{\pi} \circ G)^* \hat{g}_E = \hat{\pi}^* \hat{g}_E = \hat{\mathbf{g}}_E,$$

the star denoting pull-back.

The fixed points of G are $\mathrm{Fix}(G) = \{\mathbf{s}_1, \ldots, \mathbf{s}_n\}$, since the branch points $\mathbf{s}_l \in \mathbf{M}$ are precisely the points of the form (q, Q) with $Q = 0$.

Because G is an isometry, the pull-back G^* which is defined on $T\mathbf{M}$ restricts to a diffeomorphism on the unit tangent bundle Σ_E.

Clearly, if there are fixed points of $G^* : \Sigma_E \to \Sigma_E$, they must project to $\mathrm{Fix}(G)$. But on $T_{\mathbf{s}_l}\mathbf{M}$, G^* inverts the direction of the tangent vectors, as follows from (3.3). Thus, if one restricts G^* to the unit tangent bundle Σ_E, $\mathrm{Fix}(G^*) = \emptyset$, and by Prop. 4.1.23 of [1] the quotient $\Sigma_E/\mathbb{Z}_2$ by the action of the group $\{\mathrm{Id}, G^*\} \cong \mathbb{Z}_2$ of covering transformations is a smooth manifold.

The geodesic flow Φ_E^t is invariant w.r.t. G^* since $G : \mathbf{M} \to \mathbf{M}$ is an isometry.

To show the existence of a unique smooth extension $\pi_E : \Sigma_E \to \Sigma_E$ of $\hat{\pi}_E$ on $\hat{\Sigma}_E$, we go back to the proof of Proposition 2.3. There we introduced continuous local coordinates $(H_l, T_l, L_l, \varphi_l)$ in P to regularize the flow near the l-th nucleus. These coordinates were associated with the purely Coulombic motion, and were used to define the coordinates $(\tilde{T}_l, \tilde{L}_l, \tilde{\varphi}_l)$ adapted to the true motion. We recall that $(\tilde{H}_l, \tilde{T}_l, \tilde{L}_l, \tilde{\varphi}_l)$ were used to introduce the differential structure on the completed phase space, but the proof that the map

$$(q_1, q_2, p_1, p_2) \longleftrightarrow (\tilde{H}_l, \tilde{T}_l, \tilde{L}_l, \tilde{\varphi}_l)$$

is smooth on $\hat{U}_l^\varepsilon$ was postponed until now. To this end we use T_l, L_l and φ_l as local C^0-coordinates on the energy shell $\Sigma_E \subset P$. We are interested in the lifts of these coordinates to $T^*\mathbf{M}$ (which is isomorphic to the tangent space $T\mathbf{M}$ via the metric $\mathbf{g}_E$). In a second step, we shall control the coordinates $\tilde{T}_l$, $\tilde{L}_l$ and $\tilde{\varphi}_l$ by using the lifts of T_l, L_l and φ_l.

We assume w.l.o.g. that $\vec{s}_l = \vec{0}$. The generating function

$$S(p_1, p_2, Q_1, Q_2) := (Q_1^2 - Q_2^2)p_1 + 2Q_1Q_2p_2$$

leads to a locally canonical tranformation of contact type, supplementing the Levi-Civita transformation

$$q_1 = Q_1^2 - Q_2^2, \qquad q_2 = 2Q_1Q_2 \tag{3.9}$$

by the transformation

$$p_1 = \frac{Q_1P_1 - Q_2P_2}{2(Q_1^2 + Q_2^2)}, \qquad p_2 = \frac{Q_1P_2 + Q_2P_1}{2(Q_1^2 + Q_2^2)} \tag{3.10}$$

of the momenta.

Under this transformation the angular momentum $L_l : U_l^\varepsilon \to \mathrm{I\!R}$ relative to the l-th nucleus lifts to a function of the form

$$\mathbf{L}_l = \tfrac{1}{2}(Q_2P_1 - Q_1P_2). \tag{3.11}$$

The energy $H = \frac{1}{2}\vec{p}^{\,2} - Z_l/|\vec{q}| + W_l(\vec{q})$ transforms into

$$\mathbf{H} = (P\bar{P}/8 - Z_l)/Q\bar{Q} + W_l(Q^2). \tag{3.12}$$

The transformed Runge-Lenz vector $\mathbf{V}$ is formally undefined for $Q = 0$:

$$\mathbf{V} = -\frac{P}{8\bar{Q}}(\bar{Q}P - Q\bar{P}) - Z_l\frac{Q}{\bar{Q}}. \tag{3.13}$$

Inserting (3.12) in (3.13) with $\mathbf{H} = E$, we obtain

$$\mathbf{V} = -\frac{1}{8}P^2 + (E - W_l(Q^2))Q^2, \tag{3.14}$$

which is smooth and nonzero at $Q = 0$ for $|P|$ determined by (3.12). Furthermore, $+P$ and $-P$ lead to the same value of $\mathbf{V}$ for $Q = 0$.

We define $\pi_E : \boldsymbol{\Sigma}_E \to \Sigma_E$ by $\pi_E|_{\hat{\boldsymbol{\Sigma}}_E} := \hat{\pi}_E$, and by mapping a point $(Q = 0, P) \in \boldsymbol{\Sigma}_E$ to the point in Σ_E given by $(\tilde{H}_l, \tilde{T}_l, \tilde{L}_l, \tilde{\varphi}_l) = (E, 0, 0, \arg(\mathbf{V}))$.

Now we easily see that the time reparametrization $s : \mathbb{R} \times \boldsymbol{\Sigma}_E \to \mathbb{R}$ is smooth because, written in the local uniformizing variable $Q^2 = q$,

$$s(t, (Q_0, P_0)) = \frac{1}{\sqrt{2E}} \int_0^t \frac{Q(t')\bar{Q}(t')E\,dt'}{Z_l + Q(t')\bar{Q}(t')(E - W_l(Q^2(t')))}$$

for $(Q(t), P(t)) := \boldsymbol{\Phi}_E^t(Q_0, P_0)$. The transformation property (3.8) of the flows follows from (3.4).

In the proof of Proposition 2.3 we defined the phase space function T_l as the time needed to reach the pericenter of the orbit w.r.t. the l-th nucleus, that is, to intersect the hypersurface S_l defined on $\hat{U}_l^\varepsilon \subset T^*\hat{M}$ by the equation $q_1p_1 + q_2p_2 \equiv 0$ (we have set $\vec{s}_l = \vec{0}$) and then extended to the points of collision. Using (3.9) and (3.10), the lifted hypersurface is given by the equation $Q_1P_1 + Q_2P_2 \equiv 0$. The geodesic flow is transversal to this smooth surface.

Again, let (Q_0, P_0) denote a point in the energy shell $\boldsymbol{\Sigma}_E$ (with obvious abuse of notation) and let $(Q(t), P(t)) \equiv \boldsymbol{\Phi}_E^t(Q_0, P_0)$ be the geodesic flow line through (Q_0, P_0).

Then $\mathbf{T}_l \equiv \mathbf{T}_l(Q_0, P_0)$ may be defined in a neighbourhood of the surface $Q_1P_1 + Q_2P_2 = 0$ by the relation $Q_1(\mathbf{T}_l)P_1(\mathbf{T}_l) + Q_2(\mathbf{T}_l)P_2(\mathbf{T}_l) = 0$. By transversality of the flow w.r.t. that hypersurface, $\mathbf{T}_l$ is a smooth function.

Clearly the function T_l defined in the proof of Prop. (2.3) is related to $\mathbf{T}_l$ by

$$T_l(\pi_E(Q_0, P_0)) = s(\mathbf{T}_l(Q_0, P_0), (Q_0, P_0)), \tag{3.15}$$

so that $T_l \circ \pi_E$ is smooth.

(3.11), (3.13) and (3.15) together show that $\pi_E : \boldsymbol{\Sigma}_E \to \Sigma_E$ induces a smooth transformation from the (Q, P) coordinates to the $(\tilde{H}_l, \tilde{T}_l, \tilde{L}_l, \tilde{\varphi}_l)$ coordinates.

G^*-invariance of $\pi_E : \boldsymbol{\Sigma}_E \to \Sigma_E$ is immediate from its definition. Commutativity of the diagram is a consequence of the commutativity of a similar diagram where all manifolds wear hats.

As a side-effect, we have provided the proof of smoothness of the manifold P (this question had been postponed in the proof of Prop. (2.3)). Smoothness of P means smoothness of the canonical transformation

$$(q_1, q_2, p_1, p_2) \longleftrightarrow (\tilde{H}_l, \tilde{T}_l, \tilde{L}_l, \tilde{\varphi}_l)$$

on $\hat{U}_l^\varepsilon$.

For $\tilde{H}_l = H|_{U_l^\varepsilon}$ smoothness in the $(\vec{q}, \vec{p})$-coordinates follows from the definition of Coulombic potentials.

Smoothness of $\tilde{L}_l$ and $\tilde{\varphi}_l$ follows from the smoothness of the lifts (3.11), (3.14). Smoothness of $\tilde{T}_l$ on $\hat{U}_l^\varepsilon$ follows from (3.15). For energies $E \le V_{\max}$ where $\boldsymbol{\Phi}_E^t$ is undefined, we can nevertheless argue in a similar way, since the question is a local one. □

Now $(\mathbf{M}, \mathbf{g}_E)$ is a geodesically complete Riemannian manifold and we can analyse the geodesic motion on that surface. Then we find all energy E solution curves of our original problem by projecting the *E-geodesics* (the geodesics of the metric $\mathbf{g}_E$) on $\mathbf{M}$ down to $M \cong \mathbb{C}$, using π.

To a large extent, geodesic motion on $(\mathbf{M}, \mathbf{g}_E)$ is determined by the Gaussian curvature K_E of $\mathbf{g}_E$. On $\hat{\mathbf{M}}$ we may use local $\vec{q}$ coordinates. Then one finds

$$K_E(\vec{q}) = \frac{(1 - V(\vec{q})/E)\Delta V(\vec{q})/E + (\nabla V(\vec{q})/E)^2}{2(1 - V(\vec{q})/E)^3} \tag{3.16}$$

with Δ and ∇ denoting the Euclidean Laplacian and gradient, respectively.

For the case of a single Coulomb potential of the form $V(\vec{q}) = -Z/|\vec{q}|$ the Gaussian curvature equals

$$K_E(\vec{q}) = -\frac{Z}{2E(|\vec{q}| + Z/E)^3}. \tag{3.17}$$

We begin with some elementary estimates concerning the dependence of the Gaussian curvature $K_E(q)$ on q and E. For $l \in \{1, \ldots, n\}$ let

$$\mathbf{U}_l(r) := \{(q, Q) \in \mathbf{M} \mid |q - s_l| < r\}$$

be a (small) disk of Euclidean radius $r > 0$ and center $\mathbf{s}_l$.

At this point we have a word of warning concerning our notation for Euclidean distances: Even if $q \in \mathbf{M}$ denotes a generic point of $\mathbf{M}$, we write $|q|$ for the Euclidean norm of the projected point $\pi(q) \in M = \mathbb{C} = \mathbb{R}^2$.

Lemma 3.2

1. *There exist $r > 0$ and $C_1 > 0$ such that for all sufficiently large $E > V_{\max}$*

$$|K_E(q)| < C_1/E \text{ for all } q \in \mathbf{M} \setminus \bigcup_{l=1}^{n} \mathbf{U}_l(r)$$

and

$$|K_E(q)| < \frac{C_1}{E} |q|^{-2-\epsilon} \text{ for } |q| > R_{\min}, \tag{3.18}$$

whereas

$$K_E(q) < 0 \text{ for all } q \in \bigcup_{l=1}^{n} \mathbf{U}_l(r).$$

2. *There exist $C_2, C_3 > 0$ such that for all sufficiently large $E > V_{\max}$*

$$K_E(q) < -C_2 \text{ for all } q \in \bigcup_{l=1}^{n} \mathbf{U}_l(C_3 E^{-1/3}).$$

3. *For $q \in \bigcup_{l=1}^{n} \mathbf{U}_l(r)$, $K_E(q) = \mathcal{O}(E^2)$.*

Proof. We write $V(\vec{q}) = f_l(\vec{q})/|\vec{q} - \vec{s}_l|$. So f_l is bounded together with its derivatives near $\vec{s}_l$. From (2.3) we obtain

$$\lim_{\vec{q}\to\vec{s}_l} f_l(\vec{q}) = -Z_l < 0. \tag{3.19}$$

Rewriting (3.16) using the functions f_l, we get

$$K_E = \frac{[f_l - 2(\vec{q}-\vec{s}_l)\nabla f_l + |\vec{q}-\vec{s}_l|^2 \Delta f_l]/E + |\vec{q}-\vec{s}_l|\,[(\nabla f_l)^2 - f_l \Delta f_l]/E^2}{2(|\vec{q}-\vec{s}_l| - f_l/E)^3}. \tag{3.20}$$

For E larger than a fixed positive constant and $|\vec{q}-\vec{s}_l|$ small enough, the leading term in the numerator of (3.20) is $f_l(\vec{q})/E$ which is negative by (3.19), showing the third part of statement 1. The first part concerning the absolute value of $K_E(q)$ follows by inspection of (3.16).

By (2.4) and (2.5), the Gaussian curvature K_E meets an estimate of the form (3.18) that is, the manifold $(\mathbf{M}, \mathbf{g}_E)$ is asymptotically flat.

Statement 2 holds since for $|\vec{q}-\vec{s}_l| < C_3 E^{-1/3}$ the denominator in (3.20) can be estimated by E^{-1} from below. Statement 3 follows from the local behaviour near a nucleus which can be read off from (3.17).

Statement 3 follows by inspection of (3.20). □

Denoting the surface element of $\mathbf{g}_E$ by $d\mathbf{M}_E$, eq. (3.18) implies that the total curvature $\int_{\mathbf{M}} K_E(\vec{q}) d\mathbf{M}_E$ is finite.

Lemma 3.3 *For a Coulombic potential V with n nuclei, the total curvature of $(\mathbf{M}, \mathbf{g}_E)$ equals*

$$\int_{\mathbf{M}} K_E(\vec{q}) d\mathbf{M}_E = -2\pi \cdot n,$$

independently of the energy $E > V_{\mathbf{max}}$.

Proof. We show that $\int_{\hat{M}} K_E(\vec{q}) d\hat{M}_E = -\pi \cdot n$ (with obvious notation), using the Gauss-Bonnet Theorem. Then the Lemma follows, since $\hat{\mathbf{M}}$ is a two-fold covering of $\hat{M}$ and since $\mathbf{g}_E$ is a smooth metric so that $\int_{\hat{\mathbf{M}}} K_E(\vec{q}) d\mathbf{M}_E = \int_{\mathbf{M}} K_E(\vec{q}) d\mathbf{M}_E$.

To apply the Gauss-Bonnet Theorem, we integrate the geodesic curvature of a family $\gamma_r : S^1 \to U_l$ of small loops parametrized by $r > 0$, where $U_l \subset \hat{M}$ is a neighbourhood of the position $\vec{s}_l$ of the l-th nucleus. Using Euclidean polar coordinates (r, φ) centered at $\vec{s}_l$, we set $\gamma_r(t) := (r, t)$. Then

$$\hat{g}_E((r,\varphi)) = (1 - V((r,\varphi))/E)(dr \otimes dr + r^2 d\varphi \otimes d\varphi),$$

and we calculate the geodesic curvature

$$k_g(\gamma_r)(t) := \frac{\vec{e}_2(t) \cdot (\nabla \vec{e}_1(t)/dt)}{|\dot{\gamma}_r(t)|}$$

with the Frenet frame

$$\begin{aligned} \vec{e}_1(t) &:= \left(r\sqrt{1 - V((r,t))/E}\right)^{-1} \frac{\partial}{\partial \varphi} \\ \vec{e}_2(t) &:= \left(\sqrt{1 - V((r,t))/E}\right)^{-1} \frac{\partial}{\partial r} \end{aligned}$$

and $\nabla \vec{e}_1(t)/dt := \nabla_{\dot\gamma_r(t)}\vec{e}_1(t)$ denoting the covariant derivative of $\vec{e}_1$ along $\gamma_r(t)$ (see, e.g. Klingenberg [25]).

With obvious notation for the Christoffel symbols we obtain

$$\vec{e}_2(t)\cdot(\nabla\vec{e}_1(t)/dt) = \frac{1}{r}\Gamma^r_{\varphi\varphi}(\gamma_r(t))$$

by using the expression

$$\frac{\nabla X(\gamma_r(t))}{dt} = \left(\frac{d}{dt}(X^i(\gamma_r(t))) + \Gamma^i_{jk}(\gamma_r(t))X^j(\gamma_r(t))\dot\gamma^k_r(t)\right)\frac{\partial}{\partial x^i}$$

for the covariant derivative of a vector field $X = X^i\partial/\partial x^i$. Since $|\dot\gamma_r(t)| = r\sqrt{1-V((r,t))/E}$, the geodesic curvature is given by

$$k_g(\gamma_r)(t) = \frac{\Gamma^r_{\varphi\varphi}(\gamma_r(t))}{r^2\sqrt{1-V((r,t))/E}}.$$

Writing $f(r,t) := V((r,t))\cdot r^\alpha$, $\alpha\in\mathbb{R}$,

$$\Gamma^r_{\varphi\varphi}(\gamma_r(t)) = -\frac{g_{\varphi\varphi,r}(\gamma_r(t))}{2g_{rr}(\gamma_r(t))} = -\frac{r^{1-\alpha}(2Er^\alpha - (2-\alpha)f(r,t) - r\partial f(r,t)/\partial r)}{2(E-f(r,t)r^{-\alpha})}.$$

Thus the total geodesic curvature

$$I_r := \int_{\gamma_r(S^1)} k_g(\gamma_r)\cdot ds = \int_0^{2\pi} k_g(\gamma_r)(t) r\sqrt{1-V((r,t))/E}dt$$

of the loops equals

$$I_r = \int_0^{2\pi}\frac{-2E + (2-\alpha)r^{-\alpha}f(r,t) + r^{1-\alpha}\partial f(r,t)/\partial r)}{2(E-f(r,t)r^{-\alpha})}dt. \tag{3.21}$$

Setting $\alpha := 1$ and using the limiting behaviour of V near $\vec{s}_l$ described in the definition of Coulombic potentials, we get

$$\lim_{r\searrow 0} I_r = -\pi,$$

that is, half of the value one would get without a Coulombic singularity. Using the same formula (3.21) to control the geodesic curvature of a large circular loop encircling all singularities once, we obtain (with $\alpha := 0$)

$$\lim_{r\to\infty} I_r = -2\pi,$$

the total geodesic curvature of a circle in the Euclidean plane.

Thus every nucleus contributes a term $-(2\pi - \pi)$ to the total Gaussian curvature $\int_{\hat M} K_E(\vec{q})d\hat{M}_E$, using Gauss-Bonnet. □

As the energy E becomes large, the Gaussian curvature goes to zero pointwise on $(\hat{\mathbf{M}}, \hat{\mathbf{g}}_E)$, but not uniformly. Instead, the curvature concentrates near

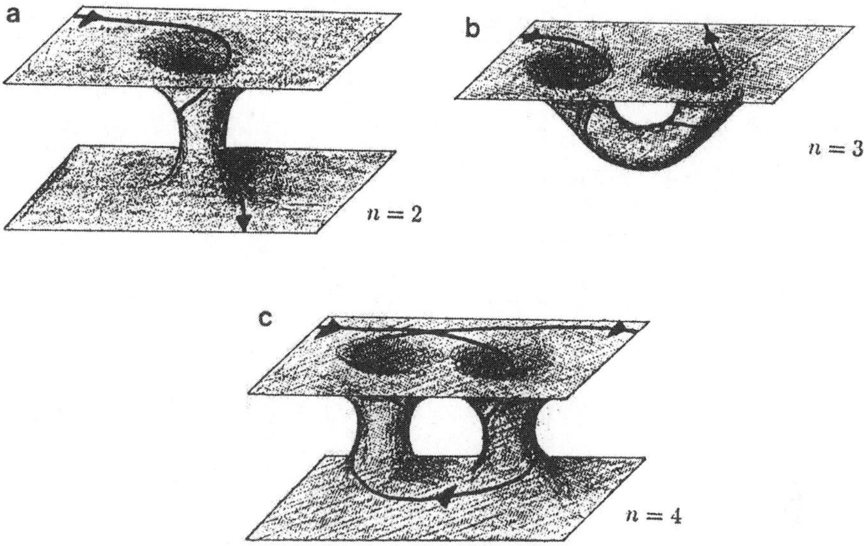

Fig. 3.1. The surfaces $(\mathbf{M}, \mathbf{g}_E)$ for a) $n = 2$, b) $n = 3$, c) $n = 4$ with typical geodesics.

the branch points $\mathbf{s}_l$ of the covering $\pi : \mathbf{M} \to M$. This is to be expected since it is only near the nuclei that a high-energy particle is strongly deflected. Next we consider the topological type of the surface $\mathbf{M}$. We denote by S_g a compact orientable surface of genus g.

Lemma 3.4 *If the number of nuclei is even ($n = 2m$), then* $\mathbf{M}$ *is homeomorphic to* $S_{m-1} \setminus \{i_1, i_2\}$.

If the number of nuclei is odd ($n = 2m - 1$), then $\mathbf{M}$ *is homeomorphic to* $S_{m-1} \setminus \{i_1\}$.

Proof. We use the defining equation (3.2) for the Riemann surface in a slightly different form. A compactification $\bar{\mathbf{M}}$ of $\mathbf{M}$ is given as the concrete Riemann surface defined by

$$Q^2 = \prod_{l=1}^{n}(q - s_l), \quad (q, Q) \in \mathbb{P} \times \mathbb{P},$$

$\mathbb{P} = \mathbb{C} \cup \{\infty\}$ denoting the Riemann sphere. $\bar{\mathbf{M}}$ is a branched covering of $\mathbb{P}$. For $n = 2m - 1$ as well as for $n = 2m$, the total branching number B equals $2m$, but infinity is a branch point only in the odd case. The covering is $k = 2$-sheeted. The Riemann-Hurwitz formula allows to calculate the genus $\mathcal{G}(\bar{\mathbf{M}})$ of $\bar{\mathbf{M}}$ (see, e.g., Farkas and Kra [11]):

$$\begin{aligned} \mathcal{G}(\bar{\mathbf{M}}) &= \frac{B}{2} + k(\mathcal{G}(\mathbb{P}) - 1) + 1 \\ &= m + 2(0 - 1) + 1 = m - 1. \end{aligned}$$

$\bar{\mathbf{M}}$ has two points projecting to $\infty \in \mathbb{P}$ if n is even and one such point if n is odd. □

Thus we have

- $n = 1$, $\mathbf{M} \cong \mathbb{C}$ (a plane)
- $n = 2$, $\mathbf{M} \cong \mathbb{R} \times S^1$ (a cylinder)
- $n = 3$, $\mathbf{M} \cong \mathbb{T}^2 \setminus \{i_1\}$ (a torus with one point deleted)
- $n = 4$, $\mathbf{M} \cong \mathbb{T}^2 \setminus \{i_1, i_2\}$ (a torus with two points deleted).

Fig. 3.1 shows models of $\mathbf{M}$ which preserve some aspects of the asymptotically flat metric $\mathbf{g}_E$.

For a qualitative analysis of the bounded orbits we need to determine the fundamental group $\pi_1(\mathbf{M})$ of $\mathbf{M}$.

Lemma 3.5 *For n nuclei, the fundamental group $\pi_1(\mathbf{M})$ is isomorphic to the free group on $n-1$ generators.*

Thus, in particular, $\pi_1(\mathbf{M}) \cong \{\mathrm{Id}\}$ for $n = 1$, $\pi_1(\mathbf{M}) \cong \mathbb{Z}$ for $n = 2$, and $\pi_1(\mathbf{M})$ is non-abelian for $n > 2$.

Proof. We use the representation of compact surfaces by their fundamental polygons. For $n = 1$ and $n = 2$ the statement is obviously true, since $\mathbf{M} \cong \mathbb{C}$, $\mathbf{M} \cong \mathbb{R} \times S^1$, respectively. So let $n \geq 3$, that is, $g \geq 1$. The fundamental polygon of S_g has $4g$ edges which are pairwise identified, so that we can denote them by $a_1, b_1, \ldots, a_g, b_g$. The fundamental relation is

$$a_1 b_1 a_1^{-1} b_1^{-1} \cdots a_g b_g a_g^{-1} b_g^{-1} = \mathrm{id},$$

see Fig. 3.2.

First we consider the case $n = 2m - 1$, $g = m - 1$. $D := S_g \setminus (a_1 \cup \cdots \cup b_g)$ is an open disk. If one point $i_1 \in D$ is deleted, then the deformation retract of $S_g \setminus \{i_1\}$ is a $2g$-leaved bouquet, with $2g = 2(m-1) = n - 1$. So $\pi_1(\mathbf{M})$ is isomorphic to the free group on $n - 1$ generators.

The case $n = 2m$, $g = m - 1$ is similar. We use an additional based loop e which meets the edges $a_1, \ldots, b_{m-1}$ only in the base point and which connects opposite points in the polygon. Then $S_g \setminus (e \cup a_1 \cup \cdots \cup b_g)$ consists of two open disks D_1, D_2. If we delete $i_1 \in D_1$ and $i_2 \in D_2$ from S_g, then the bouquet consisting of $e \cup a_1 \cup \cdots \cup b_g$ is a deformation retract of that space. Again, that bouquet consists of $2(m-1) + 1 = n - 1$ leaves. □

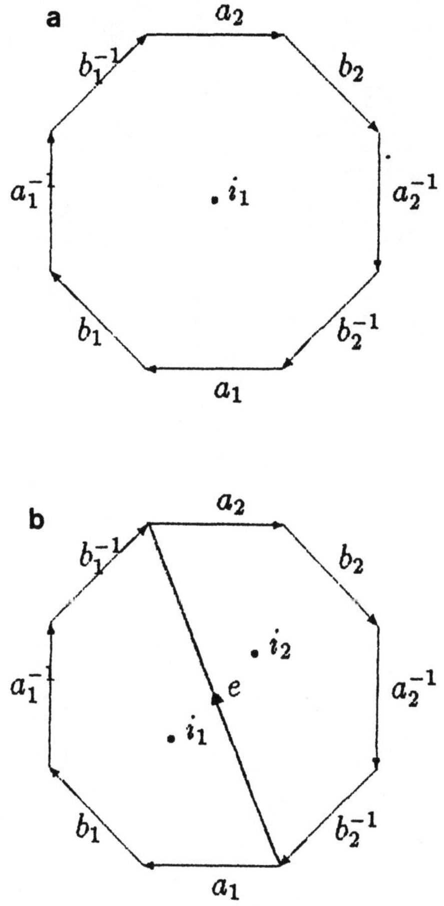

Fig. 3.2. The fundamental polygons of the surfaces a) $S_g \setminus \{i_1\}$, b) $S_g \setminus \{i_1, i_2\}$ for $g = 2$.

4. Periodic Orbits

In this chapter we are to analyse the positive energy periodic orbits. The basic technique is to minimize the energy functional on the infinite-dimensional manifold of H^1-curves $c : I \to \mathbf{M}$, with $I := [0,1]$. As we shall see, this approach gives us full qualitative information for E large.

We begin by introducing some notation and quoting results from Klingenberg [26] and Schwartz [36]. If $E > V_{\max}$, then $(\mathbf{M}, \mathbf{g}_E)$ is a complete Riemannian manifold. We first discuss the general setting of an arbitrary complete Riemannian manifold (M, g).

Absolutely continuous curves $c : I \to M$ with finite *energy functional*

$$\mathcal{E}(c) := \tfrac{1}{2} \int_0^1 g_{c(t)}(\dot{c}(t), \dot{c}(t))dt \tag{4.1}$$

are called *H^1-curves*, and we denote by $H^1(I, M)$ the set of these curves. For $c \in H^1(I, M)$ the *length*

$$\mathcal{L}(c) := \int_0^1 \left(g_{c(t)}(\dot{c}(t), \dot{c}(t))\right)^{1/2} dt \tag{4.2}$$

is well-defined and finite.

By Thm. 2.3.12 of [26], $H^1(I, M)$ is a Hilbert manifold, using variational vector fields along a curve $c \in H^1(I, M)$ to define a chart near c.

Furthermore, by Thm. 2.3.19 of [26], $H^1(I, M)$ possesses a Riemannian metric $\| \cdot \|_1 : TH^1(I, M) \to \mathbb{R}$ induced by the Riemannian inner product on the tangent space $T_cH^1(I, M)$ at $c \in H^1(I, M)$, which is given by

$$\langle \xi_1, \xi_2 \rangle_1 := \int_0^1 g_{c(t)}(\xi_1(t), \xi_2(t))dt + \int_0^1 g_{c(t)}(\nabla \xi_1(t), \nabla \xi_2(t))dt, \tag{4.3}$$

for $\xi_1, \xi_2 \in T_cH^1(I, M)$, $\nabla \xi_i(t)$ denoting the covariant derivative of ξ_i along $c(t)$.

Then the energy functional $\mathcal{E} : H^1(I, M) \to \mathbb{R}$ is a C^1-function, with derivative

$$d\mathcal{E}(c)(\xi) = \int_0^1 g_{c(t)}(\dot{c}(t), \nabla \xi(t))dt$$

(Thm. 2.3.20).

We shall be interested in geodesic segments c with fixed end points $c(0) = p$ and $c(1) = q$, $p, q \in \mathbf{M}$, and in closed geodesics in $\mathbf{M}$. Therefore, we consider certain submanifolds $\Omega_{pq}M$ and ΛM of $H^1(I, M)$ which are defined as follows:

Let $\Pi : H^1(I, M) \to M \times M$ be given by $\Pi(c) := (c(0), c(1))$, that is, Π projects to the end points of the curves c.

Then one defines the space $\Omega_{pq}M$ by

$$\Omega_{pq}M := \Pi^{-1}(\{(p,q)\}),$$

and the space ΛM of parametrised closed curves (loops) by

$$\Lambda M := \Pi^{-1}(\Delta), \text{ with } \Delta := \{(p,q) \in M \times M \mid p = q\}$$

denoting the diagonal.

$\Omega_{pq}M$ and ΛM are closed submanifolds of $H^1(I, M)$, since Π is a submersion, and since these spaces are defined as the preimages of the closed submanifolds $\{(p,q)\}$ and Δ, respectively.

Our first task will be to show that $H^1(I, M)$ is a complete metric space for (M, g) complete, slightly generalizing Thm. 2.4.7 of [26]. Then $\Omega_{pq}M$ and ΛM are complete metric spaces too, since they are closed.

We denote by $\exp : TM \to M$ the *exponential map* which sends a tangent vector $X \in T_pM$ to the end point $c(1)$ of the geodesic segment $c(t)$ of length $|X| := \sqrt{g_p(X, X)}$ with $c(0) := p$, $\dot{c}(0) := X$. Since (M, g) is complete by assumption, the exponential map is well-defined, and we write $\exp_p$ for the restriction of $\exp$ to T_pM.

The infimum of the lengths of curves connecting two points $p, q \in M$ defines their distance $d_M(p, q)$. Thus, we may endow the space $C(I, M)$ of continuous curves $c : I \to M$ with a *distance function* given by

$$d(c_1, c_2) := \sup_{t \in I} d_M(c_1(t), c_2(t)).$$

Then $C(I, M)$ is a complete metric space, since (M, g) is a complete metric space (see, e.g., Prop. 1.1.15 of [1]).

On the other hand, we can derive from the Riemannian metric $\|\cdot\|_1$ on $H^1(I, M)$ another distance function $d_1 : H^1(I, M) \times H^1(I, M) \to \mathbb{R} \cup \{\infty\}$ in the following way:

Let $\chi : I \to H^1(I, M)$ be a C^1-curve. The *length* $\mathcal{L}_1(\chi)$ of χ is then given by

$$\mathcal{L}_1(\chi) := \int_0^1 \left\| \frac{d}{ds}\chi(s) \right\|_1 ds. \tag{4.4}$$

Then the distance d_1 is defined by

$$d_1(c_0, c_1) := \inf\{\mathcal{L}_1(\chi) \mid \chi : I \to H^1(I, M) \text{ is } C^1, \chi(0) = c_0, \chi(1) = c_1\}.$$

To define distances d_{pq} and d_Λ on the submanifolds $\Omega_{pq}M$ resp. ΛM of $H^1(I, M)$, one takes the infimum over curves with values in these spaces. If $c_0, c_1 \in \Omega_{pq}M$ are not homotopic, we set $d_{pq}(c_0, c_1) := \infty$, and similarly for ΛM. Clearly, $d_{pq}(c_0, c_1) \geq d_1(c_0, c_1)$, and similarly for ΛM.

The following proposition is a generalization of Thm. 2.4.7 of [26].

Proposition 4.1 *If a Riemannian manifold (M,g) is complete, then the metric space $(H^1(I,M),d_1)$ of H^1-curves is complete.*

Proof. The inclusion $H^1(I,M) \hookrightarrow C(I,M)$ is continuous, since for any C^1-path $\chi : I \to H^1(I,M)$ with end points $\chi(0) = c_0$, $\chi(1) = c_1$ we may find a $t_M \in I$ such that $d(c_0,c_1) = d_M(c_0(t_M), c_1(t_M))$. But then

$$\begin{aligned} d(c_0,c_1) &= d_M(c_0(t_M),c_1(t_M)) \leq \int_0^1 \left|\frac{\partial}{\partial s}\chi(s,t_M)\right| ds \\ &\leq \int_0^1 \max_t \left(\left|\frac{\partial}{\partial s}\chi(s,t)\right|\right) ds \leq \sqrt{2}\int_0^1 \left\|\frac{\partial \chi}{\partial s}(s)\right\|_1 ds = \sqrt{2}\mathcal{L}_1(\chi), \end{aligned}$$

using (4.4). The last inequality is a Sobolev estimate.

Now a Cauchy sequence $\{c_k\}$ in $H^1(I,M)$ converges to $c \in C(I,M)$, since $C(I,M)$ is a complete metric space. There exists a compact neighbourhood $N \subset M$ of the curve $\{c(t) \mid t \in I\}$ such that the curves $\{c_k(t) \mid t \in I\} \subset N$ for $k \geq k_0$ large enough. On compact sets the injectivity radius of the exponential map is positive (Prop. 2.1.10 of [26]). Therefore, there exists a $K \geq k_0$ and a vector field $\xi \in T_{c_K}C(I,M)$ such that $c(t) = \exp_{c_K(t)}\xi(t)$. Furthermore, for a suitable such K, there exist for all $k \geq K$ vector fields $\xi_k \in T_{c_K}H^1(I,M)$ with $c_k = \exp_{c_K}(\xi_k)$, where exp now denotes the exponential map on the Hilbert manifold $H^1(I,M)$. The ξ_k form a Cauchy sequence in $T_{c_K}H^1(I,M)$ which converges to $\xi \in T_{c_K}H^1(I,M)$, since this tangent space is a Hilbert space (Thm 8.6 of [36]). Thus c is a H^1-curve. □

To analyse the structure of the periodic orbits, we must show that the loop space under consideration meets a *Palais-Smale condition.* That is, we would like to know that topologically nontrivial closed loops in $\mathbf{M}$ of bounded energy accumulate at closed geodesics if the gradient of the energy functional goes to zero. Inspection of Fig. 3.1 shows that this fact cannot be a consequence of the topological type of the manifold alone but depends on its metric. Intuitively, the important point is that one cannot pull a nontrivial loop to infinity without letting its length (and thus the value of $\mathcal{E}$) grow without bounds. To show that, we must control the injectivity radius.

Definition 4.2 Let (M,g) be a complete Riemannian manifold and $q \in M$. The *injectivity radius of* q is the supremum $\imath(q) \in \mathbb{R}^+ \cup \{\infty\}$ of $\rho \in \mathbb{R}^+$ such that $\exp_q$, restricted to the ball $\{v \in T_qM \mid |v| < \rho\}$, is injective.

The *injectivity radius* $\imath(M)$ *of* M is defined as $\imath(M) := \inf_{q\in M}\imath(q)$.

Lemma 4.3 *If $E > V_{\max}$, then the injectivity radius $\imath(\mathbf{M})$ of $(\mathbf{M},\mathbf{g}_E)$ is strictly positive. Furthermore,*

$$\lim_{|q|\to\infty} \imath(q) = \infty. \tag{4.5}$$

For E large, $\imath(\mathbf{M}) > \frac{1}{3}d_{\min}$, with $d_{\min}$ defined in (2.7).

Proof. We show (4.5). Then $\imath(\mathbf{M}) > 0$, since the injectivity radius is strictly positive on compact sets.

By (2.24), for $|q| \geq R_{\text{vir}}(E)$, the Jacobi metric $\mathbf{g}_E$ can be uniformly controlled by the Euclidean metric, since $|V(\vec{q})/E| < \frac{1}{2}$. Furthermore, since $\mathbf{M}$ is asymptotically flat, for $|q|$ large, $|K_E(p)| < |q|^{-2-\epsilon/2}$ for all $p \in \mathbf{M}$ with $|p-q| < \frac{1}{2}|q|$, using (3.18). We show that $\imath(q) > |q|/4$. Assume that this estimate is wrong. Then there exist two geodesic segments of maximal length $|q|/4$ which both start at q and have a common end point p. Without loss of generality we assume that q and p are the only points belonging to both geodesic segments, so that they enclose a domain A of area $|A|$. Let $\alpha > 0$ be the angle enclosed at q. Then for $|q|$ large, we can estimate $|A|$ from above by $|q|^2 \cdot \alpha$. On the other hand, the total Gaussian curvature of the enclosed domain is bounded by

$$\int_A K_E(p) d\mathbf{M}_E < |A| \cdot |q|^{-2-\epsilon/2}$$

so that $\int_A K_E(p) d\mathbf{M}_E < \alpha \cdot |q|^{-\epsilon/2}$. This is in contradiction with the Gauss-Bonnet theorem for the geodesic biangle which shows the assertion.

To show that $\imath(\mathbf{M}) > \frac{1}{3}d_{\text{min}}$, we remark that for E large, $\imath(q) > \frac{2}{3}R_{\text{min}} \geq \frac{1}{3}d_{\text{min}}$ for $|q| > 2R_{\text{min}}$, using (3.18). Therefore, we assume $|q| \leq 2R_{\text{min}}$. The set $D := \{r \in \mathbf{M} \mid |r-q| < \frac{2}{5}d_{\text{min}}\}$ which projects to a disk of Euclidean radius $\frac{2}{5}d_{\text{min}}$ in the configuration plane contains at most one of the branch points $\mathbf{s}_l$, $l \in \{1, \ldots, n\}$. Therefore D is a simply connected open neighbourhood of q, and any curve c connecting q with a point outside D has length $\mathcal{L}^E(c) > \frac{1}{3}d_{\text{min}}$. Therefore we conclude that for E large, the injectivity radius $\imath(q) > \frac{1}{3}d_{\text{min}}$, using the estimate $K_E(r) < C_1/E$ valid inside D. □

Now let $\Lambda_N M \subset \Lambda M$ be the space of non-contractible closed loops. Again, we consider curves on the Riemannian surface $(\mathbf{M}, \mathbf{g}_E)$ for $E > V_{\text{max}}$.

We denote by $\mathcal{L}^E(c)$ the length (4.2) of a curve $c : I \to \mathbf{M}$ in the metric $\mathbf{g}_E$ and by $\mathcal{L}^\infty(c) := \mathcal{L}^\infty(\pi \circ c)$ the length of c w.r.t. the metric derived from the Euclidean metric on the base space M. This notation is consistent since $\hat{\mathbf{g}}_E = \hat{\pi}^* \hat{g}_E$ on $\hat{\mathbf{M}}$ and since $\hat{g}_E$ given by (3.1) converges pointwise to the flat metric for $E \nearrow \infty$.

Lemma 4.4 *The Palais-Smale condition holds for $\Omega_{pq}\mathbf{M}$ and $\Lambda_N\mathbf{M}$. That is, whenever we have a sequence $\{c_k\}$ in $\Omega_{pq}\mathbf{M}$ (resp. in $\Lambda_N\mathbf{M}$) such that the sequence $\{\mathcal{E}(c_k)\}$ is bounded and* $\lim_{k\to\infty} \|\text{grad}\, \mathcal{E}(c_k)\|_1 = 0$ *(where* grad $\mathcal{E}$ *denotes the gradient of $\mathcal{E}$ restricted to $\Omega_{pq}\mathbf{M}$ or $\Lambda_N\mathbf{M}$, respectively), then there exists a convergent subsequence.*

Remark 4.5 Clearly, no Palais-Smale condition holds on the component $\Lambda_0\mathbf{M} := \Lambda\mathbf{M} \setminus \Lambda_N\mathbf{M}$. For example, take a sequence $\{c_k\}$ of constant curves $c_k(t) \equiv q_k \in \mathbf{M}$ with $\lim_{k\to\infty} d_{\mathbf{M}}(q_0, q_k) = \infty$.

Proof. For $\Omega_{pq}\mathbf{M}$ this is Thm. 2.4.9 of [26]. So we consider a sequence $\{c_k\}$ of loops in $\Lambda_N\mathbf{M}$ with $\mathcal{E}(c_k) < \mathcal{E}_0$. The length of these curves is bounded: $\mathcal{L}^E(c_k) < \sqrt{2\mathcal{E}_0}$.

We claim that all these curves are then contained in a compact set of the form $\{q \in \mathbf{M} \mid |q| \leq r\}$ for a suitable $r > R_{\min}$. In fact, assume the opposite, namely the existence of a sequence of points $p_i := c_{k_i}(t_i)$ with $|p_i| \to \infty$. Then for i large enough, the curve c_{k_i} may be written as $c_{k_i}(t) = \exp_{p_i}(\xi_i(t))$ with $|\xi_i(t))| < \imath(p_i)$, using Lemma 4.3. But then the curve c_{k_i} is contractible to the constant loop $\tilde{p}_i(t) \equiv p_i$, using the homotopy $H : [0,1] \to \Lambda\mathbf{M}$ given by

$$H(a)(t) := \exp_{p_i}(a\xi_i(t)).$$

Therefore $c_{k_i} \in \Lambda_0\mathbf{M}$, contrary to our assumptions.

So we know that the curves c_k are contained in a compact subset of $\mathbf{M}$ and the proof of the Palais-Smale condition proceeds as in the proof of Thm. 8.41 of [36]. □

Theorem 4.6 *Let $E > V_{\max}$. Then on $(\mathbf{M}, \mathbf{g}_E)$, for every nontrivial homotopy class $[g] \in \pi_1(\mathbf{M})$, $[g] \neq \mathrm{Id}$, there exists a closed geodesic $c : \mathrm{I\!R}/\mathbb{Z} \to \mathbf{M}$ with c freely homotopic to g. (So closed geodesics corresponding to different conjugacy classes in $\pi_1(\mathbf{M})$ are geometrically different).*

Proof. It is well-known that the connected components of the loop space ΛM are in one-to-one correspondence with the conjugacy classes in the fundamental group $\pi_1(M)$.

The injectivity radius of $(\mathbf{M}, \mathbf{g}_E)$ is strictly positive. Therefore, the length functional $\mathcal{L}$, restricted to the space $\Lambda_N(\mathbf{M})$ of noncontractible loops, is bounded from below by a strictly positive constant. To find a closed geodesic in a component of $\Lambda_N(\mathbf{M})$, we integrate the gradient flow

$$\dot{c}(s) = -\mathrm{grad}\ \mathcal{E}(c(s)), \quad c(0) = c_0$$

to shorten a loop c_0 in that component, and use the Palais-Smale condition to show the existence of a limiting geodesic, as in [36]. □

5. Existence of an Invariant Cone Field

After this existence result for closed orbits, we consider the question of uniqueness. How many closed orbits do exist in a given conjugacy class?

The answer will be that, loosely speaking, for small positive energies there may exist many such orbits whereas for high energies there exists only one. The borderline between the two different regimes depends on the precise form of the potential. For our simplest example (2.6) of purely Coulombic potentials the threshold is at $E = 0$, and the qualitative properties of *all* positive energy bounded orbits can be fully described by our methods.

As it will turn out, the closed geodesics are unique in the above sense if they are unstable and contain no conjugate points. We already know from (2.23) that for $E > V_{\max}$ the bounded geodesics do not leave the *interaction zone*

$$\mathbf{G}_E := \{q \in \mathbf{M} \mid |q| \leq R_{\text{vir}}(E)\} \tag{5.1}$$

in positive or negative time. $\mathbf{G}_E$ is a submanifold with boundary.

Next we will show the existence of a threshold $E_1 \geq V_{\max}$ such that no geodesic segments contained in $\mathbf{G}_E$ have conjugate points for energies $E > E_1$.

The physical idea behind that statement is that, although there may be points $q \in \mathbf{G}_E$ where the Gaussian curvature $K_E(q)$ of the metric $\mathbf{g}_E$ is positive, $K_E(q)$ goes to zero pointwise as the energy E goes to infinity, if $q \neq \mathbf{s}_l$ for $l \in \{1, \ldots, n\}$, *i.e.*, if q does not project to the position $\vec{s}_l$ of a nucleus. On the other hand, long geodesic segments staying in $\mathbf{G}_E$ must come near the points $\mathbf{s}_l$ repeatedly, since only there they are deflected from a path leaving $\mathbf{G}_E$ within a short time. But the Gaussian curvature near the points $\mathbf{s}_l$ is negative, leading to instability of these orbits.

Proposition 5.1 *There exists $C_4 > 0$ and an energy $E_1 \geq V_{\max}$ such that all energy-E geodesic segments $c : I \to \mathbf{M}$ with $E > E_1$ which stay in the region*

$$\mathbf{G}_E \setminus \bigcup_{l=1}^{n} \mathbf{U}_l(C_4 E^{-1}) \tag{5.2}$$

have lengths $\mathcal{L}^E(c), \mathcal{L}^\infty(c) < 3R_{\text{vir}}(E)$.

Remark 5.2 One immediately deduces that these geodesic segments have no conjugate points if E_1 is large enough, using Thm. 2.6.2 of [26], since by Lemma 3.2 Gaussian curvature in $\mathbf{G}_E$ is bounded from above by $K_E(q) < C_1/E$.

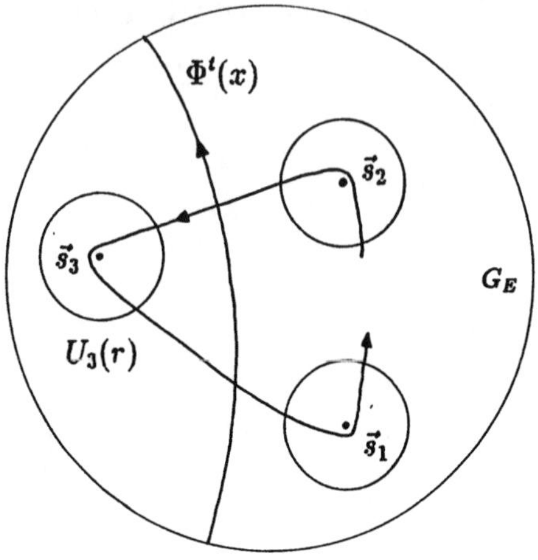

Fig. 5.1. Bounded and unbounded orbits in the original configuration plane.

Furthermore, we see that the high-energy bounded orbits reenter the region $\cup_{l=1}^{n} \mathbf{U}_l(C_4 E^{-1})$ again and again in uniformly bounded time. This follows by partitioning such an orbit into pieces of lengths $3R_{\text{vir}}(E)$.

Fig. 5.1 shows the projection of the region (5.2) to the original configuration plane.

We cut down the proof of the above proposition into three parts. The starting point is the trivial observation that a geodesic segment must have left the region $\mathbf{G}_E$ of Euclidean diameter $2R_{\text{vir}}(E)$ if the Euclidean distance between its end points $c(0)$ and $c(1)$ is larger than $2R_{\text{vir}}(E)$. We then show that this is the case if the Euclidean length of the geodesic segment is strictly larger than $2R_{\text{vir}}(E)$ and if the integrated absolute value of its geodesic curvature $k_g(c)$ (w.r.t. the Euclidean metric) is small enough. Then we estimate that geodesic curvature for the geodesics near (of distance $\leq E^{-1/3}$) the nuclei and in the complement of these regions.

Lemma 5.3 *Let $c : I \to M$ be a smooth curve in the Euclidean plane $(M, \langle \cdot, \cdot \rangle)$, parametrized proportionally to its arc length. Then*

$$|c(1) - c(0)| \geq \mathcal{L}^\infty(c) \cdot \left(1 - \int_0^1 |k_g(c)(t)||\dot{c}(s)|ds\right).$$

Proof.

$$\begin{aligned} c(1) - c(0) &= \int_0^1 \left(\int_0^s \ddot{c}(t)dt + \dot{c}(0) \right) ds \\ &= \dot{c}(0) + \int_0^1 \int_0^s \frac{\langle J\dot{c}(t), \ddot{c}(t) \rangle}{|\dot{c}(t)|^3} |\dot{c}(t)| \; J\dot{c}(t)dtds \end{aligned}$$

$$= \mathcal{L}^\infty(c)\left(\dot c(0)/|\dot c(0)| + J\int_0^1\int_0^s k_g(c)(t)\cdot\dot c(t)dtds\right)$$

with rotation matrix $J := \left(\begin{smallmatrix}0 & -1\\ 1 & 0\end{smallmatrix}\right)$, using $|\dot c(t)| \equiv \mathcal{L}^\infty(c)$ and

$$\langle \dot c(t), \ddot c(t)\rangle = \tfrac{1}{2}\tfrac{d}{dt}|\dot c|^2 = 0.$$

The estimate follows by taking absolute values. □

Next we compare the Euclidean length $\mathcal{L}^\infty(c)$ of an E-geodesic segment

$$c : I \to \mathbf{G}_E \setminus \bigcup_{l=1}^n \mathbf{U}_l(C_3E^{-1/3})$$

with its length $\mathcal{L}^E(c)$ w.r.t. the metric $\mathbf{g}_E$, and estimate its (Euclidean) geodesic curvature $k_g(c)(t)$ for high energies E.

Lemma 5.4 *As* $E \nearrow \infty$,

$$\left|\mathcal{L}^\infty(c)/\mathcal{L}^E(c) - 1\right| = \mathcal{O}(E^{-2/3}) \tag{5.3}$$

and

$$|k_g(c)(t)| = \mathcal{O}(E^{-1/3}) \quad , \quad t \in I. \tag{5.4}$$

Proof. (5.3) follows from the uniform estimate $-CE^{1/3} < V(q) < C$ for some $C > 0$ with $q \in \mathbf{G}_E \setminus \cup_{l=1}^n \mathbf{U}_l(C_3E^{-1/3})$.

To derive the estimate for the geodesic curvature, we observe that, away from the $\mathbf{s}_l \in \mathbf{M}$, the lifts π^*x_1, π^*x_2 of the canonical coordinates x_1, x_2 in M are local coordinates in $\mathbf{M}$, which we denote by x_1, x_2 for simplicity. These coordinates are well adapted to our problem since the associated Christoffel symbols of the lifted *Euclidean* metric vanish.

In these coordinates $c(t) = (x_1(t), x_2(t))$, and $c(t)$ meets the geodesic equation

$$\ddot x^i(t) + \Gamma^i_{jk}(c(t))\dot x^j(t)\dot x^k(t) = 0 \tag{5.5}$$

with Christoffel symbols

$$\Gamma^i_{jk}(x) = (2(E - V(x))^{-1}\left(\delta_{ik}\frac{\partial V}{\partial x^j} + \delta_{ji}\frac{\partial V}{\partial x^k} - \delta_{jk}\frac{\partial V}{\partial x^i}\right).$$

Thus for $q \in \mathbf{G}_E \setminus \cup_{l=1}^n \mathbf{U}_l(C_3E^{-1/3})$, the Christoffel symbols can be estimated by

$$\left|\Gamma^i_{jk}(x)\right| = \mathcal{O}(E^{-1/3}).$$

Then (5.4) follows from the equation

$$k_g(c)(t) = \frac{\langle J\dot c(t), \ddot c(t)\rangle}{|\dot c(t)|^3}$$

for the (Euclidean) geodesic curvature. □

Now we estimate the Euclidean geodesic curvature of E-geodesic segments contained in a neighbourhood $\mathbf{U}_l(C_3E^{-1/3})$ of the l-th nucleus.

Lemma 5.5 *For every $\Delta > 0$ there exists a $C_4 > 0$ such that all E-geodesic segments*

$$c : I \to \mathbf{U}_l(C_3E^{-1/3}) \setminus \mathbf{U}_l(C_4E^{-1})$$

in an annulus centered at $\hat{s}_l$ have total Euclidean geodesic curvature

$$\int_0^1 |k_g(c)(t)|\,|\dot{c}(t)|\,dt < \Delta. \tag{5.6}$$

Both their Euclidean length $\mathcal{L}^\infty(c)$ and their E-length $\mathcal{L}^E(c)$ are of order $\mathcal{O}(E^{-1/3})$ as $E \nearrow \infty$.

Proof. We start by considering Keplerian motion in the Coulomb potential $-Z_l/|\vec{q} - \vec{s}_l|$. Then we show that the perturbation due to the additional potential $W_l(\vec{q})$ defined in (2.13) vanishes as $E \nearrow \infty$.

Using polar coordinates centered at $\vec{s}_l$, the Kepler hyperbola can be written in the form of the focal equation

$$r(\varphi) = r_{\min}\frac{1+e}{1+e\cos(\varphi-\varphi_0)}$$

with eccentricity $e > 1$ and minimal distance $r_{\min}$ from the nucleus,

$$r_{\min} = (e-1)Z_l/2E.$$

The change in the direction of the particle during its passage through the annulus is bounded by $\Delta\varphi := 2\arcsin(1/e)$, so that $\Delta\varphi < \pi Z_l/(2C_4 + Z_l)$ for $r_{\min} > C_4/E$. Thus the lemma holds for purely Coulombic potentials.

Now we show the estimate for the true potential $V(\vec{q}) = -Z_l/|\vec{q} - \vec{s}_l| + W_l(\vec{q})$. The Hamiltonian equations are

$$\begin{aligned}\frac{d}{dt}\vec{p} &= -Z_l\frac{\vec{q}-\vec{s}_l}{|\vec{q}-\vec{s}_l|^3} - \nabla W_l(\vec{q})\\ \frac{d}{dt}\vec{q} &= \vec{p}.\end{aligned}$$

We rescale these equations, using the energy E of the particle, by setting $\vec{Q} := E^{1/3}(\vec{q} - \vec{s}_l)$, $\vec{P} := E^{-1/6}\vec{p}$ and $T := E^{1/2}t$. Then, in the parametrization with time T, we have

$$\begin{aligned}\frac{d}{dT}\vec{P} &= -Z_l\frac{\vec{Q}}{|\vec{Q}|^3} - E^{-2/3}\nabla_{\vec{q}}W_l(\vec{s}_l + E^{-1/3}\vec{Q})\\ \frac{d}{dT}\vec{Q} &= \vec{P}.\end{aligned}$$

The rescaled distance $\left|\vec{Q}\right|$ from the nucleus meets the bounds

$$C_4 E^{-2/3} \leq \left|\vec{Q}\right| < C_3.$$

The maximal time spent by the particle in that annulus is bounded by a constant which is E-independent in the parametrization by T if the potential is purely Coulombic.

On the other hand, the strength of the additional perturbative vector field due to the potential W_l is of the order $\mathcal{O}(E^{-2/3})$.

Therefore, we conclude that an estimate of the form (5.6) holds for the motion in the potential V, too.

The statement on the length of the geodesic segment then follows from the scaling of $\vec{Q}$ in E. □

Proof of Proposition 5.1. We demand $E_1 \geq 4(C_3/d_{\min})^3$. So we know from the outset that the different regions $\mathbf{U}_l(C_3E^{-1/3})$ do not overlap.

First we show that for E large, an E-geodesic segment of lengths $\mathcal{L}^\infty(\mathbf{c})$, $\mathcal{L}^E(\mathbf{c}) > 3R_{\text{vir}}(E)$ must intersect one of the neighbourhoods $\mathbf{U}_l(C_3E^{-1/3})$, and that it cannot reenter the neighbourhood with index $l = l_0$ without intersecting another neighbourhood with index $l_1 \neq l_0$ in the meantime.

The first assertion is a direct consequence of Lemma 5.3 and Lemma 5.4, since for an E-geodesic $\mathbf{c} : I \to \mathbf{G}_E \setminus \bigcup_{l=1}^n \mathbf{U}_l(C_3E^{-1/3})$ of lengths $\mathcal{L}^\infty(\mathbf{c}), \mathcal{L}^E(\mathbf{c}) > 2R_{\text{vir}}(E)$ projecting to $c := \pi(\mathbf{c})$, the Euclidean distance $|c(1) - c(0)|$ is larger than the Euclidean diameter $2R_{\text{vir}}(E)$ of $\mathbf{G}_E$ if E is large.

The second assertion follows from the following observation. Assume that we have an E-geodesic segment $\mathbf{c} : I \to \mathbf{G}_E \setminus \bigcup_{l=1}^n \mathbf{U}_l(C_3E^{-1/3})$ parametrized proportionally to arc length leaving the l_0-neighbourhood at time zero (that is, $\mathbf{c}(0) \in \partial\overline{\mathbf{U}}_{l_0}(C_3E^{-1/3})$ and $\dot{r}(0) \geq 0$ for $r(t) := |c(t) - \mathbf{s}_{l_0}|$). Then

$$\dot{r}(t) = \frac{\langle \dot{c}(t), c(t) - \mathbf{s}_{l_0}\rangle}{|c(t) - \mathbf{s}_{l_0}|},$$

and for E large we estimate the numerator by writing

$$\begin{aligned}
&\langle \dot{c}(t), c(t) - \mathbf{s}_{l_0}\rangle \\
&= \left\langle \dot{c}(0) + \int_0^t \ddot{c}(s)ds, c(0) - \mathbf{s}_{l_0} + t\dot{c}(0) + \int_0^t\int_0^s \ddot{c}(r)dr\, ds\right\rangle \\
&= \langle \dot{c}(0), c(0) - \mathbf{s}_{l_0}\rangle + t\,\langle \dot{c}(0), \dot{c}(0)\rangle \\
&\quad + \left\langle \int_0^t \ddot{c}(s)ds, c(0) - \mathbf{s}_{l_0} + t\dot{c}(0) + \int_0^t\int_0^s \ddot{c}(r)dr\, ds\right\rangle \\
&\quad + \left\langle \dot{c}(0), \int_0^t\int_0^s \ddot{c}(r)dr\, ds\right\rangle \\
&\geq \tfrac{1}{2}t\,\langle \dot{c}(0), \dot{c}(0)\rangle \geq 0 \qquad \text{for } 0 \leq t \leq 1,
\end{aligned}$$

since the first term is larger or equal to 0 by assumption, while the third and fourth terms may be estimated by $t \cdot \mathcal{O}(E^{-1/3})$, using the geodesic equation

(5.5). So the distance $r(t)$ from the l_0th nucleus is increasing for $0 \le t \le 1$ which shows the second assertion.

For E large, an E-geodesic c of length $3R_{\text{vir}}(E)$ intersects the neighbourhoods $\cup_{l=1}^{n}\mathbf{U}_l(C_3E^{-1/3})$ in at most

$$[1 + 6R_{\text{vir}}(E)/d_{\text{min}}]$$

segments, since the different neighbourhoods $\mathbf{U}_l$ are separated by a distance larger than $\frac{1}{2}d_{\text{min}}$.

Thus the total geodesic curvature of c can be estimated by

$$\int_0^1 |k_g(c)(t)||\dot{c}(t)|ds < 3R_{\text{vir}}(E)\cdot\mathcal{O}(E^{-1/3}) + [1 + 6R_{\text{vir}}(E)/d_{\text{min}}]\cdot\Delta,$$

using (5.4) and (5.6). $R_{\text{vir}}(E)$ is nonincreasing in E. Therefore, by choosing a suitable $C_4 > 0$ in Lemma 5.5 and E_1 large, we show that

$$\int_0^1 |k_g(c)(t)||\dot{c}(t)|ds < \frac{1}{6}$$

so that

$$|c(1) - c(0)| \ge \frac{5}{2}R_{\text{vir}}(E)$$

by Lemma 5.3. That is, the Euclidean distance of the end points of the E-geodesic is larger than the Euclidean diameter of $\mathbf{G}_E$, showing the proposition. □

Now by Remark 5.2 we know that those E-geodesics which remain in the interaction zone $\mathbf{G}_E \subset \mathbf{M}$ for a long time, meet the regions near the nuclei in uniformly bounded time intervals. We will use this information and the negativity of the Gaussian curvature near the $\mathbf{s}_l$ to show the instability of these orbits.

First we have to make precise the meaning of the word 'instability' in our context. In [2], Anosov analysed flows on compact manifolds assuming uniform instability, one important example being the geodesic flows on closed Riemannian manifolds of strictly negative sectional curvature. Since that time, these *Anosov flows* have been analysed in depth. Clearly, the motion on the energy shell cannot be of *Anosov type* since $\boldsymbol{\Sigma}_E$ is not compact. Instead, we will show the existence of an *invariant cone field* in the sense of Wojtkowski [48]. These cones live in the tangent space of the energy shell, and they are invariant in the sense that for positive time the transport of a cone w.r.t. the linear tangent flow is strictly contained in the local cone. To define these notions formally, we introduce some Riemannian geometry.

For $E > V_{\max}$ the tangent bundle $\boldsymbol{\eta} : T\mathbf{M} \to \mathbf{M}$ contains the unit tangent bundle $\boldsymbol{\eta}_E : \boldsymbol{\Sigma}_E \to \mathbf{M}$ of $(\mathbf{M}, \mathbf{g}_E)$, with

$$\boldsymbol{\Sigma}_E = \{X \in T\mathbf{M} \mid \mathbf{g}_E(X, X) = 1\}.$$

Then the geodesic motion on $(\mathbf{M}, \mathbf{g}_E)$ is the projection by $\boldsymbol{\eta}_E$ of a geodesic flow

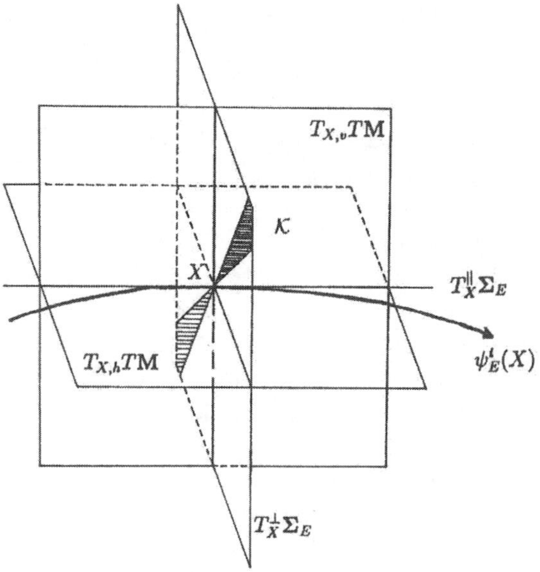

Fig. 5.2. The cone field construction

$$\boldsymbol{\Phi}_E^t : \boldsymbol{\Sigma}_E \to \boldsymbol{\Sigma}_E, \qquad t \in \mathbb{R}. \tag{5.7}$$

The tangent space $T(T\mathbf{M})$ splits into a *horizontal* and a *vertical* subspace: For $X \in T\mathbf{M}$

$$T_X T\mathbf{M} = T_{X,h}T\mathbf{M} \oplus T_{X,v}T\mathbf{M} \tag{5.8}$$

which are both canonically isomorphic to $T_q\mathbf{M}$, with $q := \boldsymbol{\eta}(X)$. Therefore, $T\mathbf{M}$ and, being a manifold, $\boldsymbol{\Sigma}_E$, carry canonical Riemannian metrics induced by $\mathbf{g}_E$.

On the other hand, by Prop. 3.2.1 of [26], the tangent bundle

$$\tau_{\boldsymbol{\Sigma}_E} : T\boldsymbol{\Sigma}_E \to \boldsymbol{\Sigma}_E \tag{5.9}$$

of $\boldsymbol{\Sigma}_E$ admits a $T\boldsymbol{\Phi}_E^t$-invariant splitting

$$\tau_{\boldsymbol{\Sigma}_E} = \tau^{\|}_{\boldsymbol{\Sigma}_E} \oplus \tau^{\perp}_{\boldsymbol{\Sigma}_E} \tag{5.10}$$

into the bundle $\tau^{\|}_{\boldsymbol{\Sigma}_E} : T^{\|}\boldsymbol{\Sigma}_E \to \boldsymbol{\Sigma}_E$ whose fibres are generated by the vector field of the flow $\boldsymbol{\Phi}_E^t$, and the bundle $\tau^{\perp}_{\boldsymbol{\Sigma}_E} : T^{\perp}\boldsymbol{\Sigma}_E \to \boldsymbol{\Sigma}_E$ of planes orthogonal to these fibres. Combining (5.8), (5.9) and (5.10), and choosing orientations, we can canonically identify each two-dimensional fibre $T_X^{\perp}\boldsymbol{\Sigma}_E$, $X \in \boldsymbol{\Sigma}_E$, with Euclidean $\mathbb{R}^2$, by mapping the unit vertical vector onto $(0,1)$ and the unit horizontal vector onto $(1,0)$. The above construction is visualized in Fig. 5.2.

A *cone* $\mathcal{K} \subset \mathbb{R}^2$ is by definition a subset of the form

$$\mathcal{K} = \mathcal{K}(\vec{v}_1, \vec{v}_2) := \{x_1\vec{v}_1 + x_2\vec{v}_2 \mid x_1, x_2 \in \mathbb{R}, x_1x_2 \geq 0\},$$

where $\vec{v}_1, \vec{v}_2 \in \mathbb{R}^2$ are linearly independent. Using the mapping

$$\mathcal{P} : \mathbb{R}^2 \setminus \{0\} \to \mathbb{R}P^1, \qquad \vec{v} \mapsto \{x\vec{v} \mid x \in \mathbb{R}\},$$

each cone is uniquely specified by an ordered pair $(\mathcal{P}(\vec{v}_1), \mathcal{P}(\vec{v}_2))$ of elements in the oriented projective space $\mathbb{R}P^1$.

For purpose of computations, we identify $\mathbb{R}P^1$ with the circle $\mathbb{R} \cup \{\infty\}$ via the map $\vec{v} \mapsto \tan(v_1/v_2)$.

Thus we shall henceforth denote a cone also by an ordered pair (S_u, S_l), where $S_u \neq S_l$ belong to $\mathbb{R} \cup \infty$.

Moreover, one can say that a cone (S_u^1, S_l^1) is *contained* in another cone (S_u^2, S_l^2) if $(S_u^2, S_u^1, S_l^1, S_l^2)$ are in cyclic order on the circle $\mathbb{R} \cup \infty$ (then the first cone is a subset of the second cone in $\mathbb{R}^2$) and that it is *strictly contained* if in addition $S_u^2 \neq S_u^1$, $S_l^2 \neq S_l^1$.

The flow $\boldsymbol{\Phi}_E^t$ on $\boldsymbol{\Sigma}_E$ induces a linear flow on $T^\perp \boldsymbol{\Sigma}_E$ which maps cones into cones. This justifies the following

Definition 5.6 A *cone field* on a subset $U \subset \boldsymbol{\Sigma}_E$ of the energy shell is a nonintersecting pair (S_u, S_l) of continuous sections S_u, S_l on the subset U in the trivial circle bundle $\tau_c : \mathbb{R}P^1 \times \boldsymbol{\Sigma}_E \to \boldsymbol{\Sigma}_E$. (We recall that points in $\mathbb{R}P^1$ are – via the above identification – identified with lines in $T^\perp \boldsymbol{\Sigma}_E$). We denote the flow on that circle bundle induced by $T\boldsymbol{\Phi}_E^t$ by ρ_E^t.

A cone field is called *invariant* if for all $t > 0$, all $x \in U$ with $\boldsymbol{\Phi}_E^t(x) \in U$, the transported cones $(\rho_E^t(S_u(x)), \rho_E^t(S_l(x)))$ are contained in the local cones $(S_u(\boldsymbol{\Phi}_E^t(x)), S_l(\boldsymbol{\Phi}_E^t(x)))$. It is called *strictly invariant* if there exists a time $\mathbf{T}_{\text{vir}} > 0$, such that the first cones are strictly contained in the second ones if $t \geq \mathbf{T}_{\text{vir}}$.

We shall only consider cone fields which never turn vertical so that we can specify cones by pairs (S_u, S_l) of reals, with $S_u > S_l$. Then the action of the flow ρ_E^t on the circle bundle can be calculated by solving the o.d.e.

$$\frac{d}{dt}S(t) + S^2(t) + K(t) = 0, \qquad S(0) := S_{u,l}^E(x), K(t) := K_E(\boldsymbol{\eta}_E(\boldsymbol{\Phi}_E^t(x))) \quad (5.11)$$

of Riccati type. This can be seen, e.g., by inspection of Thm. 3.2.17 of [26].

Furthermore, we are only interested in cone fields in the closed subregion

$$\mathbf{U}_E := \boldsymbol{\eta}_E^{-1}(\mathbf{G}_E)$$

of the energy shell projecting to the interaction zone, since only there the phenomenon of recurring orbits occurs. This is the reason why we introduced the concept of a cone field as a local object over U as above.

It is convenient to define the *exit times* $\mathbf{T}_E^\pm : \mathbf{U}_E \to \mathbb{R} \cup \{\infty\}$ to be

$$\mathbf{T}_E^\pm(x) := \sup \left\{ t \in \mathbb{R} \mid \boldsymbol{\Phi}_E^{\pm t}(x) \in \mathbf{U}_E \right\} \quad (5.12)$$

and $\mathbf{T}_{\text{vir}} := 6R_{\text{vir}}(E)$.

Proposition 5.7 *There exist positive constants s_u, s_l, I_u, I_l such that for large energies E there exist strictly invariant cone fields (S_u^E, S_l^E) on $\mathbf{U}_E$ with*

$$s_l < S_l^E(x) < S_u^E(x) < s_u \cdot E, \text{ all } x \in \mathbf{U}_E \tag{5.13}$$

such that for all $x \in \mathbf{U}_E$ and all times T, $\mathbf{T}_{\text{vir}} \le T \le \mathbf{T}_E^+(x)$

$$T I_l \ln E < \int_0^T S_l^E(\boldsymbol{\Phi}_E^t(x))dt < \int_0^T S_u^E(\boldsymbol{\Phi}_E^t(x))dt < T I_u \ln E. \tag{5.14}$$

Remarks 5.8

1. The dependence of the bounds on E is optimal.
2. We are interested in the time integrals (5.14) since these expressions will control the Lyapunov exponents.

We prove the above assertion using two basic techniques.

Firstly, we start with a simple, x-independent but non-invariant cone field and make it invariant by considering the set-theoretic union of the time-translated cones, using eq. (5.11). Due to the large negative Gaussian curvature near the nuclei, the functions S_u^E, S_l^E will become large over these regions.

Secondly, since eq. (5.11) is of first order, we can bound the true solution by comparing with the well-known solution of the Riccati equation with K constant (and independent of the energy).

Near the nuclei, we compare with the solutions of the Riccati equation for the Jacobi metric of a purely Coulombic potential:

Lemma 5.9 *The Kepler solutions of the geodesic equation in $\mathbb{R}^2$ with Jacobi metric $\sqrt{1 + Z/rE}\delta_{ik}$, $r := |\vec{q}\,|$, can be parametrized in the form*

$$r(\varphi) = \frac{L^2}{Z(1 + e\cos(\varphi - \varphi_0))}$$

with eccentricity $e := \sqrt{1 + 2EL^2Z^{-2}}$ and angular momentum L. The solutions of the Riccati equation (5.11) along these orbits have the form

$$S(r) = -\frac{E}{Z(E + \frac{Z}{r})} \frac{\sqrt{E(E + \frac{Z}{r} - \frac{L^2}{2r^2})}\cos\alpha + (E + \frac{Z}{2r})\sin\alpha}{(\frac{Er}{Z} + \frac{EL^2}{Z^2} + 1)\cos\alpha + \frac{r}{Z}\sqrt{E(E + \frac{Z}{r} - \frac{L^2}{2r^2})}\sin\alpha}, \tag{5.15}$$

with positive roots for the incoming solution and negative roots for the outgoing solution.

Proof. The parametrized solution of the Kepler equation are well-known. Clearly, they are also solutions of the geodesic equation in the Jacobi metric.

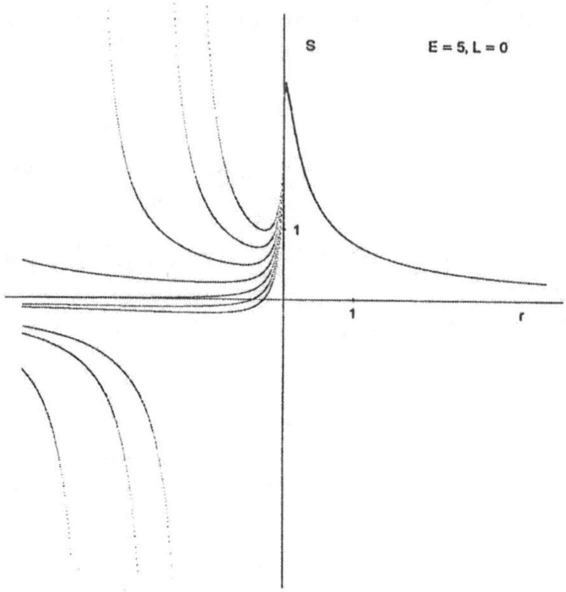

Fig. 5.3. Solutions (5.15) of the Riccati equation for the Coulomb potential, with $Z = 1$, $L = 0$ and $E = 5$. The abscissa is proportional to the radius r, with negative values for the incoming segment of the solution.

$S(r)$ defined in (5.15) solves the Riccati equation (5.11) with curvature given in eq. (3.17), as one verifies by a direct but somewhat lenghty computation, observing that the change to the parametrization by arc length t leads to a factor

$$\frac{dr}{dt} = \pm \frac{\sqrt{E(E + Z/r - L^2/2r^2)}}{E + Z/r}. \quad \square \tag{5.16}$$

We shall use Lemma 5.9 to bound the solutions of (5.11) for segments of the geodesics which enter the regions $\mathbf{U}_l(C_4 E^{-1})$. These geodesics lead to values of L^2 which are of order $\mathcal{O}(E^{-1})$, since

$$r_{\min} = \frac{Z}{2E}\left(-1 + \sqrt{1 + 2EL^2Z^{-2}}\right).$$

For these small values of the angular momentum, the solutions (5.15) resemble the $L = 0$ solutions shown in Fig. 5.3. So if S is positive before collision, then it increases sharply at the moment of collision to a value of order E. After collision, S decays approximately like the inverse distance from the nucleus, and the form of the decay does not depend much from the value of S before the collision. More precisely, we have

Lemma 5.10 *Consider eq. (5.15) for values $L^2 = \mathcal{O}(E^{-1})$ of the angular momentum and for values of the parameter α which lead to $S(r_0) > 0$ and $S(r_0) = \mathcal{O}(E^0)$ for an E-independent $r_0 > 0$ and the incoming solutions. We*

parametrize the solutions by arc length t, such that $t := 0$ for $r = r_{\min}$ and that $t \leq 0$ for the incoming solution.

Then with $r(t_0) = r_0$, we have $S(r(t)) > 0$ for $t \geq t_0$,

$$\frac{1}{2r(t)} < S(r(t)) < \frac{2}{r(t)} \qquad \text{for } t > c/E > 0 \tag{5.17}$$

and

$$\tfrac{1}{2}\ln(Er_0/Z) < \textstyle\int_{-t_0}^{t_0} S(r(t))dt < 2\ln(Er_0/Z) \tag{5.18}$$

for E large.

Proof. The positivity of $S(r(t))$ for $t \geq t_0$ follows from negativity of the curvature (3.17) for the Coulombic potential.

W.l.o.g. we assume $-\pi/2 < \alpha \leq \pi/2$ in (5.15). If $S(r_0) > 0$, then α must be smaller then zero.

Using $u := Z/Er$ and $a := L^2EZ^{-2}$, we have

$$S(r) = -\frac{Eu}{Z(1+u)}\frac{\sqrt{1+u-au^2/2}\cos\alpha + (1+u/2)\sin\alpha}{(1+(1+a)u)\cos\alpha + \sqrt{1+u-au^2/2}\sin\alpha}.$$

Expanding w.r.t. the parameter u (which is small for $r = r_0$ and E large), we obtain

$$S(r) = -\frac{Eu}{Z(1+u)}\frac{\pm(1+\mathcal{O}(u^2))\cos\alpha + \sin\alpha}{(1+(\frac{1}{2}+a)u+\mathcal{O}(u^2))\cos\alpha \pm (1+\mathcal{O}(u^2))\sin\alpha}, \tag{5.19}$$

with positive sign for the incoming solutions.

For $u = u(r_0)$ the sign of the numerator can only be different from the sign of the denominator if

$$\alpha = \frac{\pi}{4} + \mathcal{O}(u). \tag{5.20}$$

To prove (5.18), we write in obvious notation

$$\int_{-t_0}^{t_0} S(r(t))dt = \int_{r_{\min}}^{r_0} S_+(r)|dt/dr|\,dr + \int_{r_{\min}}^{r_0} S_-(r)|dt/dr|\,dr, \tag{5.21}$$

with dr/dt given by (5.16).

The first term, *i.e.*, the contribution of the incoming solution, is of order one. This follows from the estimate

$$S(r(t)) \leq \max\left(S(r(t_0)), \sqrt{-K_E(\vec{r}(t))}\right) \qquad \text{for } t_0 \leq t \leq 0$$

with curvature K_E given in (3.17) which follows from monotonicity of K_E in the radius r.

For $t > 0$ the roots have negative sign. Then (5.17) follows from (5.19) and (5.20), because for u small we have

$$S(r) = \frac{E}{Z}(u + \mathcal{O}(u^2)).\square$$

It is trivial to construct a strictly invariant cone field in the case $n = 1$ (scattering by one nucleus). In that case, for high energies every orbit leaves the region $\mathbf{U}_E$ within a time which is bounded by $\mathbf{T}_{\text{vir}}$. Therefore an invariant cone field can be found by starting with a constant cone field and performing the set-theoretic union of its time translates up to time $\mathbf{T}_{\text{vir}}$. That cone field is then trivially strictly invariant in the sense of the above definition 5.6.

In the general case we start with the constant cone field $(\hat{S}_u, \hat{S}_l)$ on $\mathbf{U}_E$ given by

$$\hat{S}_u(x) := 4/d_{\text{min}}, \quad \hat{S}_l(x) := 1/d_{\text{min}} \quad (\text{see } (2.7)),$$

and make it invariant by defining $(\tilde{S}_u^E, \tilde{S}_l^E)$ on $\mathbf{U}_E$ to be the set-theoretic union

$$\tilde{S}_u^E(x) := \max\left\{\rho_E^t(\hat{S}_u(\boldsymbol{\Phi}_E^{-t}(x))) \mid 0 \le t \le \min(\mathbf{T}_{\text{vir}}, \mathbf{T}_E^-(x))\right\}, \tag{5.22}$$

$$\tilde{S}_l^E(x) := \min\left\{\rho_E^t(\hat{S}_l(\boldsymbol{\Phi}_E^{-t}(x))) \mid 0 \le t \le \min(\mathbf{T}_{\text{vir}}, \mathbf{T}_E^-(x))\right\} \tag{5.23}$$

of time-translated cones.

We note that for E large the exit times $\mathbf{T}_E^\pm$ are continuous functions, since the boundary $\partial\mathbf{G}_E$ of the interaction zone is then geodesically convex. Therefore $(\tilde{S}_u^E, \tilde{S}_l^E)$ is a cone field (alternatively, this is a direct consequence of the virial theorem). Invariance is immediate only for $n = 1$ and is shown in the following lemma:

Lemma 5.11 *There exist $s_u, s_l > 0$ such that for E large $(\tilde{S}_u^E, \tilde{S}_l^E)$ is a strictly invariant cone field on $\mathbf{U}_E$ with*

$$s_l < \tilde{S}_l^E < \tilde{S}_u^E < s_u \cdot E. \tag{5.24}$$

Proof. The first estimate in (5.24) follows from comparison with the solution of the constant curvature Riccati equation for $K(t) := +C_1/E$ which is an upper bound for the curvature on $\mathbf{G}_E$ by Lemma 3.2. Similarly, the last inequality in (5.24) follows from part 3 of Lemma 3.2 which implies $K_E > -CE^2$ on $\mathbf{G}_E$.

To show strict invariance, we first treat $\tilde{S}_u^E$ and prove the following assertion: "If the exit time $\mathbf{T}_E^+(x) \ge \mathbf{T}_{\text{vir}}$ for $x \in \mathbf{U}_E$, then there exists a time $t_u(x) \in]0, \mathbf{T}_{\text{vir}}]$ such that $\rho_E^{t_u(x)}(\hat{S}_u(x)) < \hat{S}_u(\boldsymbol{\Phi}_E^{t_u(x)}(x)) \equiv 4/d_{\text{min}}$."

From that assertion we can derive the general inequality

$$\rho_E^t(\hat{S}_u(x)) < \tilde{S}_u^E(\boldsymbol{\Phi}_E^t(x)) \quad \text{for } \mathbf{T}_{\text{vir}} \le t < \mathbf{T}_E^+(x) \tag{5.25}$$

by noticing that due to the compactness of $\mathbf{U}_E$ we can assume that $t_u > \varepsilon > 0$ on $\mathbf{U}_E$ which by iteration implies the existence of a t_1, $t - \mathbf{T}_{\text{vir}} \le t_1 \le t$, with

$$\rho_E^{t_1}(\hat{S}_u(x)) < \hat{S}_u(\boldsymbol{\Phi}_E^{t_1}(x)),$$

proving (5.25) and thus strict upper invariance.

It is trivial to show the above assertion for $x \in \mathbf{U}_E$ projecting to a point $\boldsymbol{\eta}_E(x) \in \mathbf{G}_E$ far from the nuclei, say for $\boldsymbol{\eta}_E(x) \in \mathbf{G}_E \setminus \cup_{l=1}^n \mathbf{U}_l(d_{\min}/5)$, because there the absolute value of the curvature is bounded by C_1/E (Lemma 3.2) so that $\frac{d}{dt}\rho_E^t(\hat{S}_u(x)) < 0$ for $t = 0$. So we assume $\boldsymbol{\eta}_E(x) \in \mathbf{U}_{l_1}(d_{\min}/5)$. By Proposition 5.1 we know that the geodesic $\boldsymbol{\eta}_E(\boldsymbol{\Phi}_E^t(x))$ enters the annulus $\mathbf{U}_{l_1}(\frac{1}{2}d_{\min}) \setminus \mathbf{U}_{l_1}(d_{\min}/5)$ from within at a time $t_1 \geq 0$, and leaves it at time $t_u(x) > t_1$.

The global estimate $\rho_E^t(\hat{S}_u(x)) = \mathcal{O}(E)$ for $0 \leq t \leq \mathbf{T}_{\text{vir}}$ which follows from part 3 of Lemma 3.2 holds in particular for $t = t_1$. Furthermore, since the curvature in the annulus is small, $\rho_E^t(\hat{S}_u(x))$ can be bounded from above by $\rho_E^t(\hat{S}_u(x)) \leq (t - t_1)^{-1}$ for $t_1 \leq t \leq t_u$ and E large. On the other hand, the length of the geodesic segment $t_u - t_1 > \frac{1}{4}d_{\min}$ for E large, since then the geodesic and the Euclidean length of a curve nearly coincide. This shows that

$$\rho_E^{t_u}(\hat{S}_u(x)) < 4/d_{\min} = \hat{S}_u(\boldsymbol{\Phi}_E^{t_u}(x)) \tag{5.26}$$

so that we have proven the above assertion for $\tilde{S}_u^E$.

The proof of strict lower invariance of $\tilde{S}_l^E$ is similar. There one starts from the observation that a geodesic segment in $\mathbf{G}_E$ of length $\mathbf{T}_{\text{vir}}$ enters a neighbourhood $\mathbf{U}_l(C_4 E^{-1})$ of the l-th nucleus, as shown in Proposition 5.1. Then using Lemma 5.10, one shows that for $\mathbf{T}_E^+(x) \geq \mathbf{T}_{\text{vir}}$ there exists a time $t_l(x) \in]0, \mathbf{T}_{\text{vir}}]$ with

$$\rho_E^{t_l(x)}(\hat{S}_l(x)) > \hat{S}_l(\boldsymbol{\Phi}_E^{t_l(x)}(x)) \equiv 1/d_{\min}.$$

Lower strict invariance follows. □

Although our cone field $(\tilde{S}_u^E, \tilde{S}_l^E)$ is strictly invariant and meets (5.13), it does not meet the lower inequality in (5.14). Therefore, we define (S_u^E, S_l^E) on $\mathbf{U}_E$ by

$$S_{u/l}^E(x) := \rho_E^{t(x)}\left(\tilde{S}_{u/l}^E(\boldsymbol{\Phi}_E^{-t(x)}(x))\right) \quad \text{with } t(x) := \min(\mathbf{T}_{\text{vir}}, \mathbf{T}_E^-(x)). \tag{5.27}$$

Proof of Proposition 5.7. Strict invariance and (5.13) are immediate consequences of (5.27).

To show (5.14), we note that for $r_0 > 0$ small, a geodesic segment in $\mathbf{G}_E$ of length $T \geq \mathbf{T}_{\text{vir}}$ meets the set $\cup_{l=1}^n \mathbf{U}_l(r_0)$ in at most $2T/d_{\min}$ disjoint intervals, whereas it meets $\cup_{l=1}^n \mathbf{U}_l(C_4 E^{-1})$ in at least $2T/\mathbf{T}_{\text{vir}} - 1$ disjoint intervals. The upper bound is simple geometry, since the different $\mathbf{U}_l(r_0)$ have minimal distance larger than $d_{\min}/2$. The lower bound follows from Proposition 5.1.

Points $x \in \mathbf{U}_E$ with $\boldsymbol{\eta}_E(x) \in \mathbf{G}_E \setminus \cup_{l=1}^n \mathbf{U}_l(r_0)$ have values $S_l^E(x) < S_u^E(x)$ which are uniformly bounded independent of E.

So by changing constants, the linearity in T of the bounds appearing in (5.14) follows from the observation that for E large the main contributions to the integral come from close encounters with the nuclei, that is, from points on the orbit projecting to the regions $\mathbf{U}_l(r_0)$.

If $0 < r_0 < r$ with r from Lemma 3.2, the curvature on $\mathbf{U}_l(r_0)$ is negative, and we can apply Lemma 5.10 (with appropriate values of Z) to derive the logarithmic dependence on E in (5.14). □

6. Symbolic Dynamics

Our next task is to describe the high-energy bounded and unbounded orbits using symbolic dynamics based on the fundamental group $\pi_1(\mathbf{M})$.

By lifting the deformation retraction $H : M \times I \to M$,

$$H(\vec{q},t) := \begin{cases} \vec{q} & , |\vec{q}| \leq R_{\text{vir}}(E) \\ (tR_{\text{vir}}(E)/|\vec{q}| + (1-t))\vec{q} & , |\vec{q}| > R_{\text{vir}}(E) \end{cases}$$

to $\mathbf{M}$, we see that the manifold with boundary $\mathbf{G}_E$ is a deformation retract of $\mathbf{M}$, so that

$$\pi_1(\mathbf{G}_E) \cong \pi_1(\mathbf{M}).$$

In Lemma 3.5, $\pi_1(\mathbf{M})$ has been shown to be isomorphic to the free group on $n-1$ generators, by using a representation of the Riemann surface $\mathbf{M}$ as a punctured fundamental polygon.

To use that topological result in our present geometric context, we prove the following

Lemma 6.1 *For $n \geq 2$ and $E > V_{\max}$, there exist $n-1$ closed oriented geodesics $\mathbf{c}_l : \mathbb{R}/\mathbb{Z} \to \text{Int}(\mathbf{G}_E)$, $l \in \{1, \dots, n-1\}$ starting from $\mathbf{c}_l(0) = \mathbf{s}_n$ with $\mathbf{c}_l(\frac{1}{2}) = \mathbf{s}_l$ which are generators of $\pi_1(\mathbf{M}, \mathbf{s}_n)$. They are submanifolds intersecting only at $\mathbf{s}_n$.*

Proof. We explicitly construct the geodesics $\mathbf{c}_l$ using the representation (3.2) of $\mathbf{M}$ as a branched covering surface $\pi : \mathbf{M} \to \mathbb{C}$ with branch points $\mathbf{s}_l$.

Let $u_l : I \to \mathbb{C}$, $l \in \{1, \dots, n-1\}$, be nonintersecting curves from $u_l(0) := \vec{s}_n$ to $u_l(1) := \vec{s}_l$ and let $\tilde{u}_l : I \to \mathbf{M}$ be covering paths, that is,

$$\pi(\tilde{u}_l(t)) = u_l(t), \quad t \in I.$$

$\tilde{u}_l$ can be considered as a point in the space $\Omega_{\mathbf{s}_n,\mathbf{s}_l}\mathbf{M}$ of curves from $\mathbf{s}_n$ to $\mathbf{s}_l$.

By Lemma 4.4 the Palais-Smale condition holds on this space. Therefore, by shortening that curve, we obtain a homotopic minimal geodesic $\hat{c}_l \in \Omega_{\mathbf{s}_n,\mathbf{s}_l}\mathbf{M}$. We define the closed geodesic $\mathbf{c}_l$ by $\mathbf{c}_l(t) := \hat{c}_l(2t)$ for $0 \leq t \leq \frac{1}{2}$ and $\mathbf{c}_l(t) := G(\hat{c}_l(2(1-t)))$ for $\frac{1}{2}t \leq 1$, with $G : \mathbf{M} \to \mathbf{M}$ being the covering transformation $(q,Q) \mapsto (q,-Q)$. Then the $\mathbf{c}_l$ are generators of $\pi_1(\mathbf{M}, \mathbf{s}_n)$, as one observes using the construction of Rückkehrschnitte described, e.g., in Behnke and Sommer [3]. Self-intersections of the immersions $\mathbf{c}_l : \mathbb{R}/\mathbb{Z} \to \mathbf{M}$ cannot occur since then one

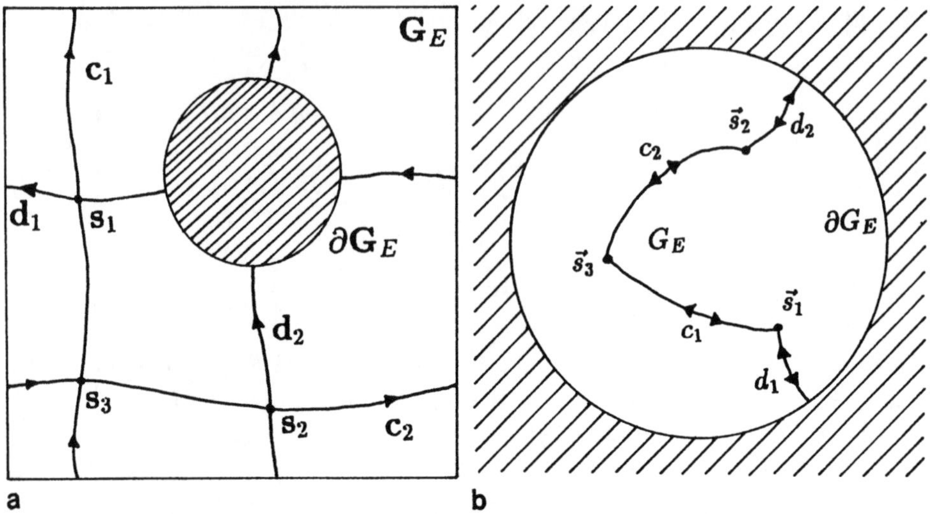

Fig. 6.1. The geodesics $\mathbf{c}_l$, $\mathbf{d}_l$ for $n = 3$ a) as subsets of $\mathbf{M}$. b) Their projections to the configuration plane

could find shorter homotopic geodesics from $\mathbf{s}_n$ to $\mathbf{s}_l$, in contradiction with the assumption of minimality. A similar argument excludes mutual intersections except at $\mathbf{c}_l(0) = \mathbf{s}_n$.

The inclusion $\mathbf{c}_l(\mathbb{R}/\mathbb{Z}) \subset \text{Int}(\mathbf{G}_E)$ is an indirect consequence of positivity of the r.h.s. of the virial identity (2.23) outside $\mathbf{G}_E$. □

We want to use the technique of Poincaré sections for doing symbolic dynamics. Therefore, we show the existence of oriented geodesics $\mathbf{d}_l : I \to \mathbf{G}_E$, which are dual to the closed geodesics $\mathbf{c}_l$ in the sense of intersection numbers, see Fig. 6.1.

Then we can attribute to an arbitrary bounded (not necessarily closed) geodesic a bi-infinite sequence of symbols which encodes the succession of transversal intersections of that geodesic with the curves $\mathbf{d}_l(I) \subset \mathbf{G}_E$.

Lemma 6.2 *For $n \geq 2$ and $E > V_{\max}$ there exist $n-1$ geodesic segments $\mathbf{d}_l : I \to \mathbf{G}_E$, $l \in \{1, \ldots, n-1\}$, whose images are neat disjoint oriented submanifolds which are perpendicular to $\partial\mathbf{G}_E$ and whose intersection numbers with the oriented geodesics $\mathbf{c}_k$, $k \in \{1, \ldots, n-1\}$ equal*

$$\#(\mathbf{d}_l, \mathbf{c}_k) = \delta_{lk}. \tag{6.1}$$

Furthermore, the minimal geodesic distance between $\mathbf{d}_k$ and $\mathbf{d}_l$, $l \neq k$, is uniformly bounded below as $E \to \infty$.

Proof. For $n \geq 2$ we claim that there exist nonintersecting curves $v_l : I \to M$, $l \in \{1, \ldots, n-1\}$, with $|v_l(0)| = R_{\text{vir}}(E)$ and $v_l(1) = \vec{s}_l$, meeting the projected geodesics $\pi \circ \mathbf{c}_k$ only for $k = l$ at $v_l(1)$. In fact, by shifting the origin of the coordinate space, if necessary, we may assume that the vectors $\vec{s}_l$, $l \in \{1, \ldots, n-1\}$, are linearly independent. We shall construct the curves $\mathbf{d}_l$ starting from the rays $v_l : I \to M$,

$$v_l(t) := \left(t + (1-t)\frac{R_{\text{vir}}(E)}{|\vec{s}_l|} \right) \cdot \vec{s}_l. \tag{6.2}$$

Let $\tilde{v}_l : I \to \mathbf{M}$ be covering paths.

Using the map $\Pi : H^1(I, \mathbf{M}) \to \mathbf{M} \times \mathbf{M}$, $\Pi(c) = (c(0), c(1))$, we define the spaces $\Omega_{\mathbf{s}_l, \partial\mathbf{G}_E}\mathbf{M}$ of curves starting at $\mathbf{s}_l$ and ending at $\partial\mathbf{G}_E$ by

$$\Omega_{\mathbf{s}_l, \partial\mathbf{G}_E}\mathbf{M} := \Pi^{-1}\left(\{\mathbf{s}_l\} \times \partial\mathbf{G}_E\right).$$

By compactness of the submanifold $\{\mathbf{s}_l\} \times \partial\mathbf{G}_E \subset \mathbf{M} \times \mathbf{M}$ it follows that $\Omega_{\mathbf{s}_l, \partial\mathbf{G}_E}\mathbf{M}$ is a closed submanifold of $H^1(I, \mathbf{M})$ meeting a Palais-Smale condition (see Chap. 4). By shortening the curves $v_l \in \Omega_{\mathbf{s}_l, \partial\mathbf{G}_E}\mathbf{M}$ we obtain closed nonintersecting geodesics $\hat{d}_l$. By the same method as in the last lemma using the covering transformation G we obtain the geodesics $\mathbf{d}_l$ which, again, must be free of mutual and self-intersections. Furthermore, $\mathbf{d}_l(\partial I) \subset \partial\mathbf{G}_E$ and they meet the boundary with a right angle because otherwise they could be shortened. The virial identity implies that $\mathbf{d}_l(]0,1[) \subset \text{Int}(\mathbf{G}_E)$, so that the $\mathbf{d}_l$ are neat (see Hirsch [18]). By uniqueness of geodesics starting with directions perpendicular to $\partial\mathbf{G}_E$ all endpoints are different.

Up to signs, the statement on the intersection numbers follows by noticing that these numbers are invariants of the deformation (see Hirsch [18], Chap. 5.2). We obtain positive signs by changing the directions of the $\mathbf{d}_l$, if necessary.

To show that the mutual geodesic distance of $\mathbf{d}_k$ and $\mathbf{d}_l$, $l \neq k$, is bounded below by $c > 0$ as $E \to \infty$, we prove that for E large, $\mathbf{d}_k(I)$ and $\mathbf{d}_l(I)$ are contained in neighbourhoods in $\mathbf{G}_E$ projecting to disjoint sectors N_k, N_l of the following form. We assume w.l.o.g. that the directions $\vec{s}_l / |\vec{s}_l|$ are ordered counterclockwise on the circle, and we consider the indices l to be integers modulo n. Then we set

$$N_l := \{\vec{q} = |\vec{q}| \exp(i\varphi) \in M \cong \mathbb{C} \mid \tfrac{1}{2}\min_k(|\vec{s}_k|) \leq |\vec{q}| \leq R_{\text{vir}}, \varphi_l^- \geq \varphi \geq \varphi_l^+\},$$

where we set $\varphi_l^{\pm} := \frac{2}{3}\varphi_l + \frac{1}{3}\varphi_{l\pm1}$ for $\vec{s}_l := |\vec{s}_l| \exp(i\varphi_l)$. So the sectors $N_l \ni \vec{s}_l$ are obtained by trisecting the angles between consecutive $\vec{s}_l$. The different sectors have positive Euclidean distance, and the distances of their lifts to $\mathbf{M}$ in the $\mathbf{g}_E$ metric is uniformly bounded below as $E \to \infty$.

Now when we shorten the path $v_l(I) \subset N_l$ in the Jacobi metric in order to obtain $\mathbf{d}_l$, we stay inside N_l if E is large, the reason being the following. Any path from $\vec{s}_l$ to the boundary $\{\vec{q} \in \mathbb{R}^2 \mid |\vec{q}| = R_{\text{vir}}\} = \pi(\partial\mathbf{G}_E)$ of the interaction zone which leaves N_l is longer than v_l in the Jacobi metric, since the minimal length of such a path in the Euclidean metric is stricly larger than $R_{\text{vir}} - |\vec{s}_l|$, using (5.3).

Thus the minimal geodesic distance between $\mathbf{d}_k$ and $\mathbf{d}_l$, $l \neq k$, is uniformly bounded below for $E \to \infty$. □

Lemma 6.3 *For E large, there exist no focal points in $\mathbf{G}_E$ and all geodesic segments $c : I \to \mathbf{G}_E$ are minimal between their end points, that is, any other curve $c' : I \to \mathbf{M}$ with $c'(0) = c(0)$ and $c'(1) = c(1)$ which is homotopic to c has geodesic length*

$$\mathcal{L}^E(c') > \mathcal{L}^E(c).$$

Proof. Existence of focal points $c(0)$, $c(1)$ along a geodesic segment c would imply existence of a non-trivial orthogonal Jacobi field $Y(t)$ along $c(t)$ with $Y(0) = \dot{Y}(1) = 0$ and $\dot{Y}(0) \neq 0 \neq Y(1)$. Then the associated solution $S(t) := \dot{Y}(t)/Y(t)$ (where the orthogonal vector field is identified with a scalar quantity) of the Riccati equation along $c(t)$ would start from $S(0) = \infty$ and decrease to $S(1) = 0$, contradicting the existence of an invariant positive cone field (S_u^E, S_l^E) shown in Proposition 5.7.

Next we show that for $\mathcal{L}^E(c') \leq \mathcal{L}^E(c)$, $c' = c$. We assume from the beginning that c' is a geodesic, too, because otherwise by shortening we could find a geodesic c'' with $\mathcal{L}^E(c'') < \mathcal{L}^E(c)$.

The non-existence of conjugate points implies that every geodesic segment from $c(0)$ to $c(1)$ has index 0. Then by the Fundamental Theorem of Morse Theory (Thm. 17.3 of Milnor [32]) the connected component of $c \in \Omega_{c(0)c(1)}\mathbf{M}$ has the homotopy type of cells of dimension 0, one for each geodesic $c' \in \Omega_{c(0)c(1)}\mathbf{M}$ homotopic to c. But such a connected complex must consist of a single point, which shows the assertion. □

Lemma 6.4 *There exist $t_{\max} > t_{\min} > 0$, $s_{\min} > 0$ such that for E large, the length of the geodesic segments $c : I \to \mathbf{G}_E$ which intersect the curves $\cup_{l=1}^{n-1}\mathbf{d}_l(I) \subset \mathbf{G}_E$ at most in the end points $c(0)$, $c(1)$, is bounded from above by*

$$\mathcal{L}^E(c) < t_{\max}, \tag{6.3}$$

and if $c(0), c(1) \in \cup_{l=1}^{n-1}\mathbf{d}_l(I)$, then

$$\mathcal{L}^E(c) > t_{\min}. \tag{6.4}$$

In the last case, the angles α_0, α_1 of intersection at the end points meet the inequality

$$|\sin \alpha_i| > s_{\min}/E. \tag{6.5}$$

If $c(0), c(1) \in \mathbf{d}_l(I)$, then $\mathrm{sign}(\alpha_0) = \mathrm{sign}(\alpha_1)$.

Proof. We begin with the last assertion. The duality (6.1) between the $\mathbf{d}_l$ and the generators $\mathbf{c}_l$ of the fundamental group implies that the open submanifold $\mathrm{Int}(\mathbf{G}_E) \setminus \cup_{l=1}^{n-1}\mathbf{d}_l(]0,1[) \subset \mathbf{M}$ is simply connected.

If c would intersect $\mathbf{d}_l$ twice with $\mathrm{sign}(\alpha_0) = -\mathrm{sign}(\alpha_1)$, then c would be homotopic to the segment of $\mathbf{d}_l(I)$ between $c(0)$ and $c(1)$, contradicting the uniqueness result of Lemma 6.3.

By transversality of $\mathbf{d}_l(I)$ and $\partial\mathbf{G}_E$, every geodesic crossing $\mathbf{d}_l(I)$ under a small angle leaves $\mathbf{G}_E \setminus \cup_{l=1}^{n-1}\mathbf{d}_l(I)$ through $\partial\mathbf{G}_E$ in both directions, showing the existence of a lower bound for $|\sin \alpha_i|$ for E fixed.

To show the scaling of (6.5) with E, we remark that if the Jacobi metric inside a strip along $\mathbf{d}_l(I)$ came from a purely Coulombic potential centered at $\mathbf{s}_l = \mathbf{d}_l(\frac{1}{2})$, then the eccentricity e of those Kepler solutions crossing $\mathbf{d}_l(I)$ with an angle $|\sin\alpha| \leq s_{\min}/E$ would be bounded uniformly in E by $1 \leq e \leq 1+\varepsilon$, and by choosing a small $s_{\min} > 0$, $\varepsilon > 0$ could be made arbitrarily small (compare with the proof of Lemma 5.5). So these solutions would leave the strip through $\partial\mathbf{G}_E$.

For a general Jacobi metric we introduce Fermi coordinates in a strip along the geodesic $\mathbf{d}_l$. Then we compare with the purely Keplerian motion.

To prove (6.4), we assume first that $c(i) \in \mathbf{d}_{l_i}(I)$, $i = 0,1$, with $l_1 \neq l_0$. The compact ∂-manifolds $\mathbf{d}_{l_0}(I)$ and $\mathbf{d}_{l_1}(I)$ are disjoint and their minimal geodesic distance is bounded from below by an E-independent positive constant.

This can be seen by inspection of (6.2), which shows that for $\varepsilon > 0$ and E large, $\mathcal{L}^E(\mathbf{d}_l) < 2(R_{\text{vir}}(E) - |\vec{s}_l|) + \varepsilon$. By Lemma 6.2 there exists an E-independent lower bound for the mutual geodesic distance of $\mathbf{d}_{l_0}(I)$ and $\mathbf{d}_{l_1}(I)$.

If $l_0 = l_1$, then by the last assertion, $\mathcal{L}^E(c) > \imath(\mathbf{M})$, the injectivity radius of $\mathbf{M}$, which by Lemma 4.3 is bounded by $\imath(\mathbf{M}) \geq \frac{1}{3}d_{\min}$.

To show (6.3), we remark that for a *fixed* large energy E' a bound $t'_{\max}$ exists.

In fact, consider the closure K of a fundamental domain of $\mathbf{G}_{E'} \setminus \cup_{l=1}^{n-1}\mathbf{d}_l(I)$ in the universal cover of $\mathbf{G}_{E'}$. K is compact and simply connected, and the length function (w.r.t. the lifted Jacobi metric for energy E') can be considered as a continuous function of the end points of a geodesic segment in K, since there is a unique geodesic segment in K connecting these end points (see Lemma 6.3). Thus the length function on $K \times K$ attains a maximum $t'_{\max}$.

We assume that $E' > 2V_{\max}$. Then for all $E > E'$ one has $\sqrt{1 - V/E} \leq \sqrt{2}\sqrt{1 - V/E'}$ which implies that $t_{\max} := \sqrt{2}t'_{\max}$ is a uniform upper bound for $\mathcal{L}^E(c)$. □

As a consequence of Lemma 6.4, one may describe the long-time behaviour of all geodesic segments in $\mathbf{G}_E$, E large, by iterating a Poincaré map defined below.

For $n = 1$, by (6.3) every geodesic leaves $\mathbf{G}_E$ after a time bounded by $t_{\max}$. Therefore, we assume $n \geq 2$.

Let $D_k := \boldsymbol{\eta}_E^{-1}(\mathbf{d}_k(I)) \subset \boldsymbol{\Sigma}_E$ be the set of points in the energy shell projecting to the geodesic $\mathbf{d}_k$, $k \in \{1,\ldots,n-1\}$. Then D_k is a two-dimensional ∂-submanifold. For $x \in \mathbf{M}$, $\boldsymbol{\eta}_E^{-1}(x)$ is diffeomorphic to a circle. Therefore, $D_k \cong I \times S^1$. In order to find coordinates on the cylinder D_k, we define the explicit diffeomorphism

$$\mathcal{I}_k : D_k \to I_k \times S^1, \qquad \text{with } I_k := \left[-\tfrac{1}{2}\mathcal{L}^E(\mathbf{d}_k), \tfrac{1}{2}\mathcal{L}^E(\mathbf{d}_k)\right]$$

by

$$(q,\dot{q}) \mapsto (l(q), \alpha(q,\dot{q})), \tag{6.6}$$

where $l(q)$ is the signed geodesic distance (on $\mathbf{d}_k$) between q and $\mathbf{s}_k = \mathbf{d}_k(\frac{1}{2})$, and $\alpha(q,\dot{q})$ is the angle between $\dot{q}$ and $\dot{\mathbf{d}}_k(t)$, with $\mathbf{d}_k(t) = q$. Note that the orientations of $\mathbf{d}_k$ and of $\mathbf{M}$ make these quantities well-defined.

The two curves on D_k with $\sin(\alpha(q,\dot{q})) = 0$ correspond to the orbit projecting to the geodesic $\mathbf{d}_k$, and its time reverse. Therefore, an orbit crossing $\mathbf{d}_k$ in positive, resp. negative direction meets D_k in the rectangles

$$C_k := \mathcal{I}_k^{-1}(\{(l,\alpha) \in \mathcal{I}_k(D_k) \mid \sin\alpha > 0\}), \tag{6.7}$$

respectively

$$C_{k+n-1} := \mathcal{I}_k^{-1}(\{(l,\alpha) \in \mathcal{I}_k(D_k) \mid \sin\alpha < 0\}). \tag{6.8}$$

Henceforth we use the coordinates $(l, \cos\alpha)$ on the rectangles C_k which will be called Poincaré sections (or surfaces). Let the *return time* $\mathbf{T}_E : \cup_{k=1}^{2(n-1)} C_k \to \mathrm{I\!R}$ be given by

$$\mathbf{T}_E(x) := \inf\left\{ t > 0 \,\middle|\, \boldsymbol{\Phi}_E^t(x) \in \bigcup_{k=1}^{2(n-1)} C_k \cup \boldsymbol{\eta}_E^{-1}(\partial \mathbf{G}_E) \right\}. \tag{6.9}$$

Then $\mathbf{T}_E < t_{\max}$, and $\mathbf{T}_E(x) > t_{\min}$ if $\boldsymbol{\Phi}_E^{\mathbf{T}_E(x)}(x) \in \cup_{k=1}^{2(n-1)} C_k$, see Lemma 6.4.

We shall analyse the *Poincaré map* $\mathbf{P}$ on $\cup_{k=1}^{2(n-1)} C_k$, given by

$$\mathbf{P}(x) := \boldsymbol{\Phi}_E^{\mathbf{T}_E(x)}(x). \tag{6.10}$$

For $k_0, k_1 \in \{1, \ldots, 2(n-1)\}$ let $W(k_0,k_1) \subset C_{k_1}$, $V(k_0,k_1) \subset C_{k_0}$ be given by

$$W(k_0,k_1) := C_{k_1} \cap \{\mathbf{P}(x) \mid x \in C_{k_0}\}, \tag{6.11}$$

and

$$V(k_0,k_1) := \{x \in C_{k_0} \mid \mathbf{P}(x) \in C_{k_1}\}. \tag{6.12}$$

Then by a transversality argument, using (6.5),

$$\mathbf{P}(k_0,k_1) : V(k_0,k_1) \to W(k_0,k_1), \qquad \mathbf{P}(k_0,k_1) := \mathbf{P}|_{V(k_0,k_1)}$$

is a diffeomorphism.

By the sign condition of Lemma 6.4 no geodesic turns back. This implies that $V(k_0,k_1) = \emptyset = W(k_0,k_1)$ for $|k_1 - k_0| = n-1$.

In the next lemma, the Poincaré map is described in more detail, see Fig. 6.2.

Lemma 6.5 *Let* $k_0, k_1 \in \{1, \ldots, 2(n-1)\}$, *with* $|k_1 - k_0| \neq n-1$.
Then there exist smooth, strictly decreasing functions

$$v_l(k_0,k_1), v_u(k_0,k_1) : I_{k_0} \to]-1,1[$$

with $v_l(k_0,k_1) < v_u(k_0,k_1)$ *and smooth, strictly increasing functions*

$$w_l(k_0,k_1), w_u(k_0,k_1) : I_{k_1} \to]-1,1[$$

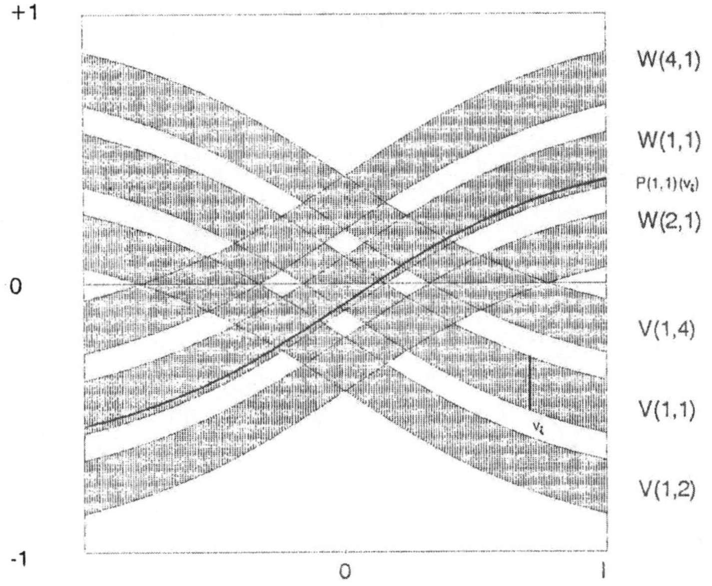

Fig. 6.2. The rectangle D_1 with the domains $V(1,k)$ and images $W(k,1)$ of the Poincaré map, for $n = 3$.

with $w_l(k_0,k_1) < w_u(k_0,k_1)$ *such that*

$$V(k_0,k_1) = \{(x,y) \in I_{k_0} \times]-1,1[\,|\, v_l(k_0,k_1)(x) \le y \le v_u(k_0,k_1)(x)\}$$

and

$$W(k_0,k_1) = \{(x,y) \in I_{k_1} \times]-1,1[\,|\, w_l(k_0,k_1)(x) \le y \le w_u(k_0,k_1)(x)\}\,.$$

The diffeomorphisms $\mathbf{P}(k_0,k_1) : V(k_0,k_1) \to W(k_0,k_1)$ *preserve the area two-form* $dl \wedge d\cos\alpha$.

For $i \in I_{k_0}$ *the image* $\mathbf{P}(k_0,k_1)(v_i) \subset W(k_0,k_1)$ *of the vertical segment* $v_i := \{(i,y) \mid v_l(k_0,k_1)(i) \le y \le v_u(k_0,k_1)(i)\}$ *is (in the sense above) a strictly increasing curve in* $W(k_0,k_1)$.

Conversely, the pre-image $(\mathbf{P}(k_0,k_1))^{-1}(w_i) \subset V(k_0,k_1)$ *of any vertical segment* $w_i := \{(i,y) \mid w_l(k_0,k_1)(i) \le y \le w_u(k_0,k_1)(i)\}$ *is (in the sense above) a strictly decreasing curve in* $V(k_0,k_1)$.

Proof. We linearize $\mathbf{P}(k_0,k_1)$ at a point $(l_0, \cos\alpha_0) \in V(k_0,k_1)$ and write this linear transformation in the form of a 2×2-matrix

$$A := \begin{pmatrix} a_{11} & a_{12} \\ a_{21} & a_{22} \end{pmatrix},$$

using the coordinate bases $(\partial/\partial l_0, \partial/\partial\cos\alpha_0)$, $(\partial/\partial l_1, \partial/\partial\cos\alpha_1)$. It is useful to write the transformation in the form

$$A = T(\alpha_1)\, U\, T^{-1}(\alpha_0), \tag{6.13}$$

where $U = (u_{ij})$ is the monodromy matrix arising as the solution of the Jacobi equation,

$$\frac{d}{dt}U(t) = \begin{pmatrix} 0 & 1 \\ -K(t) & 0 \end{pmatrix} U(t), \qquad U(0) := \mathbf{1} \tag{6.14}$$

along the geodesic segment with end points $(l_i, \alpha_i) \in C_{k_i}$, $i = 0, 1$.

The matrices $T(\alpha)$ in (6.13) transform the Jacobi coordinates $(Y, \dot{Y})$ on the space of orthogonal variations of a geodesic into the coordinates w.r.t. the bases $(\partial/\partial l, \partial/\partial \cos\alpha)$. Therefore, they have the form

$$T(\alpha) = \begin{pmatrix} (\sin\alpha)^{-1} & 0 \\ 0 & \sin\alpha \end{pmatrix}.$$

The vector field $a_{12}\partial/\partial l + a_{22}\partial/\partial \cos\alpha$ is tangent to the curve $\mathbf{P}(k_0, k_1)(v_i)$. Therefore, if $a_{12} \neq 0$ and $a_{22}/a_{12} > 0$, then this curve is strictly increasing.

We have

$$a_{12} = u_{12}/(\sin\alpha_0 \sin\alpha_1) \quad \text{and} \quad a_{22} = u_{22} \sin\alpha_1 / \sin\alpha_0.$$

a_{12} is bounded away from zero since $|\sin\alpha_i| > s_{\min}/E$ for $i = 0, 1$ by (6.5) and since $u_{12} = 0$ would imply the existence of a focal point, contradicting Lemma 6.3.

We have $a_{22}/a_{12} > 0$ since $a_{22}/a_{12} = \sin^2\alpha_1 u_{22}/u_{12}$, and since $u_{22}/u_{12} > s_l > 0$ with s_l from Proposition 5.7, as a consequence of our existence result for a cone field.

This proves that the curves $\mathbf{P}(k_0, k_1)(v_i)$ are strictly increasing.

The assertion on $w_u(k_0, k_1) > w_l(k_0, k_1)$ follows by noticing that these boundary curves are the images of the two vertical curves v_i, $i \in \partial I_{k_0}$.

$\mathbf{P}(k_0, k_1)$ is symplectic since $\det(A) = +1$. The statements on the preimage curves $\mathbf{P}(k_0, k_1)^{-1}(w_i)$, $v_u(k_0, k_1)$ and $v_l(k_0, k_1)$ follow from considerations similar to those above, using the matrix

$$A^{-1} := \begin{pmatrix} a_{22} & -a_{12} \\ -a_{21} & a_{11} \end{pmatrix},$$

and noticing that $u_{22}(t) \equiv u_{11}(t)$. □

Up to now we do not yet know whether the images and preimages of the Poincaré map intersect as suggested by Fig. 6.2. This is the content of

Lemma 6.6 *Let $k_0, k_1, k_2 \in \{1, \ldots, 2(n-1)\}$ with $|k_1 - k_0| \neq n - 1 \neq |k_2 - k_1|$. Then*

$$w_u(k_0, k_1)(i) < 0 < v_l(k_1, k_2)(i) \quad \textit{for } i \in I_{k_1} \textit{ small},$$

whereas

$$v_u(k_1, k_2)(i) < 0 < w_l(k_0, k_1)(i) \quad \textit{for } i \in I_{k_1} \textit{ large}.$$

Proof. We show the first inequality $w_u(k_0, k_1)(i) < 0$ for $i \in I_{k_1}$ small, the others being similar in character. More precisely, by continuity of w_u it clearly suffices to show that inequality for i being the lower boundary of I_{k_1}.

Assume to the opposite that $w_u(k_0, k_1)(i) \geq 0$, that is, the cosine of the angle between $\dot{\mathbf{d}}_{k_1}(0)$ and the direction of the geodesic segment starting at a boundary point of $\mathbf{d}_{k_0}(I)$ is larger then zero. We know from Lemma 6.2 that $\dot{\mathbf{d}}_{k_1}(0)$ is perpendicular to the boundary $\partial\mathbf{G}_E$ and points inside. So by our assumption our geodesic would enter $\mathbf{G}_E$ at $\mathbf{d}_{k_1}(0)$ or would be tangential to the boundary. Both possibilities are excluded by the virial inequality. □

Now we can set up symbolic dynamics by attributing to a geodesic intersecting $\mathbf{G}_E$ a sequence of symbols determined by the succession of its intersections with the curves $\mathbf{d}_k$.

For that purpose let the *symbol space* $\mathcal{S}$ be simply

$$\mathcal{S} := \{1, \ldots, 2(n-1)\}, \tag{6.15}$$

equipped with the discrete topology.

To analyse the bounded orbits we introduce the space $\{f : \mathbb{Z} \to \mathcal{S}\}$ of bi-infinite sequences, equipped with the product topology. The quadratic *transition matrix* $\mathcal{M}$ of size $2(n-1)$ is given by

$$\mathcal{M}_{kl} := \begin{cases} 0 & , \text{ for } |l-k| = n-1 \\ 1 & , \text{ otherwise.} \end{cases} \tag{6.16}$$

Then the space $\mathbf{X}$ of *admissible sequences* is given by

$$\mathbf{X} := \{f : \mathbb{Z} \to \mathcal{S} \mid \forall i \in \mathbb{Z} : \mathcal{M}_{f(i)f(i+1)} = 1\}. \tag{6.17}$$

On $\mathbf{X}$ the *shift* $\sigma : \mathbf{X} \to \mathbf{X}$, $\sigma(f)(i) := f(i+1)$, is a homeomorphism. It is well-known that the topology on $\mathbf{X}$ is generated by the metric

$$d(f,g) := \sum_{i \in \mathbb{Z}} 2^{-|i|} \cdot (1 - \delta_{f(i),g(i)}), \quad f, g \in \mathbf{X}. \tag{6.18}$$

We also use finite admissible sequences $(k_0, \ldots, k_m)$, $k_i \in \mathcal{S}$, and the spaces $\mathbf{X}^\pm$ of admissible sequences $f : \mathbb{Z}^\pm \to \mathcal{S}$, with $\mathbb{Z}^\pm := \{z \in \mathbb{Z} \mid \pm z \geq 0\}$. Again, these sequences are called admissible if for successive symbols the transition matrix equals one.

Using (6.12) and (6.11), we inductively define the sets $V(k_0, \ldots, k_m)$ for $(k_0, \ldots, k_m)$ admissible, $m > 1$, by

$$V(k_0, \ldots, k_m) := \mathbf{P}(k_0, k_1)^{-1}(W(k_0, k_1) \cap V(k_1, \ldots, k_m)).$$

Similarly, for $(k_{-m}, \ldots, k_0)$ admissible, $m > 1$, let

$$W(k_{-m}, \ldots, k_0) := \mathbf{P}(k_{-1}, k_0)(V(k_1, k_0) \cap W(k_{-m}, \ldots, k_{-1})).$$

The maps $\mathbf{P}(k_0, \ldots, k_m) : V(k_0, \ldots, k_m) \to W(k_0, \ldots, k_m)$ are then defined by concatenation of the $\mathbf{P}(k_i, k_{i+1})$ and are diffeomorphisms.

Lemma 6.7 *There exists $m_0 \in \mathbb{N}$ and $J_u > J_l > 0$ such that for E large and $(k_{-m}, \ldots, k_0)$ admissible there exist smooth, strictly increasing functions*

$$w_l(k_{-m}, \ldots, k_0), w_u(k_{-m}, \ldots, k_0) : I_{k_0} \to]-1, 1[$$

of slope

$$\begin{aligned} \frac{d}{di} w_{u/l}(k_{-m}, \ldots, k_0)(i) &> s_l \left(1 - w_{u/l}^2(k_{-m}, \ldots, k_0)(i)\right) \\ &> s_l (s_{\min}/E)^2 \text{ for } i \in I_{k_0}, \end{aligned} \tag{6.19}$$

with

$$\begin{aligned} W(k_{-m}, \ldots, k_0) &= \{(x, y) \in I_{k_0} \times]-1, 1[\, | \\ & \quad w_l(k_{-m}, \ldots, k_0)(x) \le y \le w_u(k_{-m}, \ldots, k_0)(x)\}. \end{aligned} \tag{6.20}$$

Moreover, if $m \ge m_0$, then

$$\exp(-J_u m \ln E) < w_u(k_{-m}, \ldots, k_0) - w_l(k_{-m}, \ldots, k_0) < \exp(-J_l m \ln E) \tag{6.21}$$

and

$$\frac{d}{di} w_{u/l}(k_{-m}, \ldots, k_0)(i) < s_u \cdot E \text{ for } i \in I_{k_0}. \tag{6.22}$$

Similar estimates apply to the strictly decreasing functions $v_{u/l}(k_0, \ldots, k_m) : I_{k_0} \to]-1, 1[$ whose graphs bound the strip $V(k_0, \ldots, k_m)$.

Proof. The graphs of $w_{u/l}(k_{-m}, \ldots, k_0)$ are images under $\mathbf{P}(k_{-m}, \ldots, k_0)$ of vertical segments v_i in $V(k_{-m}, \ldots, k_0)$, with $i \in \partial I_{k_{-m}}$. This follows by induction in m, using the fact that the intersections $W(k_{l-1}, k_l) \cap V(k_l, k_{l+1})$ of the domains and images of the Poincaré maps are diffeomorphic to rectangles, as a consequence of Lemma 6.5 and 6.6. This proves (6.20).

Estimate (6.19) is a consequence of the existence of an invariant cone field with E-independent lower bound $s_l > 0$, see (5.13). Similar to the proof of Lemma 6.5, we have

$$\frac{d}{di} w_{u/l}(k_{-m}, \ldots, k_0)(i) = \sin^2(\alpha_0(i))\, u_{22}(i)/u_{12}(i), \tag{6.23}$$

with $u_{22}(i)/u_{12}(i) > s_l > 0$.

By definition of $w_{u/l}$, one has $\cos(\alpha_0(i)) = w_{u/l}(k_{-m}, \ldots, k_0)(i)$. Therefore,

$$\sin^2(\alpha_0(i)) = 1 - w_{u/l}^2(k_{-m}, \ldots, k_0)(i) > (s_{\min}/E)^2,$$

using (6.5).

Instead of proving (6.21) we shall prove the analogous bound for $v_{u/l}$ which reads:

$$\exp(-J_u m \ln E) < v_u(k_0, \ldots, k_m) - v_l(k_0, \ldots, k_m) < \exp(-J_l m \ln E)$$

We first show the existence of $m_0' \in \mathbb{N}$ such that for E large, upon $m \geq m_0'$ iterations using the monodromy matrix U (see (6.13)), the image of any vertical vector lies in an invariant cone field.

W.l.o.g. we assume in Lemma 6.4 that $t_{\max} \geq 2\mathbf{T}_{\text{vir}}$, with $\mathbf{T}_{\text{vir}} = 6R_{\text{vir}}(E)$. Consider the solution $S(t)$ of the Riccati equation along an orbit segment $\boldsymbol{\Phi}_E^t(x)$ of length $2\mathbf{T}_{\text{vir}}$ in $\mathbf{U}_E$, with $S(0) := \infty$. We then claim that $S(2\mathbf{T}_{\text{vir}})$ is contained in the local invariant cone, that is,

$$S_l^E(\boldsymbol{\Phi}_E^{2\mathbf{T}_{\text{vir}}}(x)) < S(2\mathbf{T}_{\text{vir}}) \leq S_u^E(\boldsymbol{\Phi}_E^{2\mathbf{T}_{\text{vir}}}(x)). \tag{6.24}$$

In view of (5.27), assertion (6.24) follows if we show that

$$S(\mathbf{T}_{\text{vir}}) \leq \tilde{S}_u^E(\boldsymbol{\Phi}_E^{\mathbf{T}_{\text{vir}}}(x)), \tag{6.25}$$

with $\tilde{S}_u^E$ defined in (5.22). The proof of (6.25) is nearly verbally the same as the proof of (5.26) in Lemma 5.11.

Now let $m_0' := [t_{\max}/t_{\min}] + 1$, so that after $m > m_0'$ iterations the tangents to the image $\mathbf{P}(k_0, \ldots, k_m)(v_i)$ of the vertical curve

$$v_i := \{(i, y) \mid v_l(k_0, \ldots, k_m)(i) \leq y \leq v_u(k_0, \ldots, k_m)(i)\}$$

$i \in I_{k_0}$ are contained in the local cones. We use the abbreviation

$$\Delta := v_u(k_0, \ldots, k_m)(i) - v_l(k_0, \ldots, k_m)(i),$$

and want to prove that Δ is exponentially small in m by considering the iterates of the curve v_i. The image $\mathbf{P}(k_0, \ldots, k_{m_0'})(v_i)$ of this curve is then a strictly increasing curve whose horizontal length Δ' can be bounded by

$$\tfrac{1}{2}\Delta m_0' t_{\min} \leq \Delta' \leq \tfrac{\Delta}{s_l} \exp(m_0' t_{\max} I_u \ln E) \cdot (E/s_{\min})^2 \tag{6.26}$$

for E large.

These estimates are found by inspection of the entry

$$a_{12} = u_{12}/(\sin \alpha_0 \sin \alpha_{m_0'}) \tag{6.27}$$

in the linearization A of the Poincaré map $\mathbf{P}(k_0, \ldots, k_{m_0'})$. The entry $u_{12}(t)$ of the matrix $U(t)$ (see (6.14)) meets the differential equation $\ddot{u}_{12}(t) = -K(t)u_{12}(t)$ with initial conditions $u_{12}(0) = 0$, $\dot{u}_{12}(0) = 1$.

The lower bound in (6.26) follows from the estimate $K(t) < C_1/E$ (part one of Lemma 3.2) which implies

$$u_{12}(t) > \sqrt{E/C_1} \sin\left(\sqrt{C_1/E}\,t\right) > \tfrac{1}{2}t \quad \text{for } 0 < t < \tfrac{\pi}{2}\sqrt{E/C_1}.$$

The upper bound in (6.26) is similar to the upper estimate in (5.14). The factor $(E/s_{\min})^2$ comes from a lower bound for the denominator in the relation (6.27), using (6.5).

After $m - m_0'$ more iterations, the image $\mathbf{P}(k_0, \ldots, k_m)(v_i)$ of the vertical curve v_i has horizontal length Δ'' bounded by

$$\Delta k_l \exp\left(m t_{\min} I_l \ln(E)\right) < \Delta'' < \Delta k_u \exp(m t_{\max} I_u \ln(E)), \qquad (6.28)$$

with $k_u := 1/s_l$ and

$$k_l := \tfrac{1}{2} m_0' t_{\min} \exp\left(-m_0' t_{\min} I_l \ln(E)\right).$$

Now notice that we know that the horizontal length Δ'' after m iterations equals the length of the interval I_{k_m}. That length is bounded from above and from below by E-independent constants.

We now solve the inequalities in (6.28) for Δ. We notice that the form of the dependence of k_u and k_l on E allows us to absorb these coefficients in the exponentials by choosing suitable constants

$$J_u \geq t_{\max} I_u \quad \text{and} \quad J_l \leq t_{\min} I_l$$

if $m \geq m_0$ for $m_0 := 2m_0'$.

The upper bound (6.22) on the slope of the functions $w_{u/l}$ follows from (6.23) since $u_{22}(i)/u_{12}(i) < s_u \cdot E$ for $m \geq m_0$ by strict invariance of the cone field and (5.13). □

For an admissible sequence $f^+ = (k_0, k_1, \ldots) \in \mathbf{X}^+$ we define

$$V(f^+) := \bigcap_{m \in \mathbb{N}} V(k_0, \ldots, k_m) \subset C_{k_0}. \qquad (6.29)$$

Similarly, for an admissible sequence $f^- = (\ldots, k_{-1}, k_0) \in \mathbf{X}^-$ we define

$$W(f^-) := \bigcap_{m \in \mathbb{N}} W(k_{-m}, \ldots, k_0) \subset C_{k_0}. \qquad (6.30)$$

We note that these sets are nested: $V(k_0, \ldots, k_{m_2}) \subset V(k_0, \ldots, k_{m_1})$ for $m_2 \geq m_1$. Furthermore, by estimate (6.21), the vertical diameter of these sets goes to zero as $m \to \infty$ and by (6.19) and (6.22) they are bounded by curves with a common Lipschitz constant. Thus $V(f^+)$ is a continuous strictly decreasing curve whereas $W(f^-)$ is continuous and strictly increasing.

Let

$$\Lambda^+ := \bigcup_{f^+ \in \mathbf{X}^+} V(f^+), \quad \Lambda^- := \bigcup_{f^- \in \mathbf{X}^-} W(f^-) \qquad (6.31)$$

and $\Lambda := \Lambda^+ \cap \Lambda^-$.

By restriction, we associate half-infinite admissible sequences $f^\pm \in \mathbf{X}^\pm$ to $f \in \mathbf{X}$. Then we define a map $\mathcal{H} : \mathbf{X} \to \Lambda$ by

$$\mathcal{H}(f) := V(f^+) \cap W(f^-).$$

Note that in view of Lemma 6.6 the curves $V(f^+)$ and $W(f^-)$ in C_{k_0} intersect. Since $V(f^+)$ is strictly decreasing whereas $W(f^-)$ is strictly increasing, their intersection consists of precisely one point, which we identify with an element of Λ.

Furthermore, the metric on $\boldsymbol{\Sigma}_E$ induced by $\mathbf{g}_E$ induces a distance

$$\text{dist} : \boldsymbol{\Sigma}_E \times \boldsymbol{\Sigma}_E \to \mathbb{R}$$

which we restrict to Λ.

Lemma 6.8 *There exist $\alpha > 0$, $C > 0$ such that for E large, $\mathcal{H}$ is an α-Hölder continuous homeomorphism, that is,*

$$\text{dist}(\mathcal{H}(f), \mathcal{H}(g)) \leq C d^{\alpha}(f,g), \qquad \textit{all } f,g \in \mathbf{X},$$

conjugating the shift with the Poincaré map $\mathbf{P}_\Lambda := \mathbf{P}|_\Lambda$:

$$\mathcal{H} \circ \sigma = \mathbf{P}_\Lambda \circ \mathcal{H}. \tag{6.32}$$

Proof. $\mathcal{H}$ is injective since

$$V(k_0, \ldots, k_m) \cap V(k'_0, \ldots, k'_m) = \emptyset \quad \text{for } (k'_0, \ldots, k'_m) \neq (k_0, \ldots, k_m)$$

and a similar property for the W strips.

Continuity of $\mathcal{H}$ is a consequence of (6.21), (6.19) and (6.22) which imply the exponential estimate in m

$$|l_2 - l_1| + |\cos\alpha_2 - \cos\alpha_1| \leq 2\exp(-J_l m \ln E)(E/s_{\min})^2 s_l^{-1} \cdot (1 + s_u E) \tag{6.33}$$

for two points $(l_1, \cos\alpha_1), (l_2, \cos\alpha_2) \in V(k_0, \ldots, k_m) \cap W(k_{-m}, \ldots, k_0)$, valid for $m \geq m_0$. Note that in view of (6.5) $\text{dist}((l_1, \cos\alpha_1), (l_2, \cos\alpha_2))$ is equivalent to the l.h.s. of (6.33)

$\mathcal{H}^{-1}$ is continuous since Λ is compact.

Symbol sequences $f, g \in \mathbf{X}$ with $f(i) = g(i) = k_i$ for $|i| \leq m$ have distance $d(f,g) \leq 2^{1-m}$, see (6.18). Therefore, the existence of a Hölder exponent α valid for all large E follows from (6.33). The constant $C > 0$ may be chosen E-independent since the diameter of the set Λ is bounded from above by an E-independent constant. □

For $n \geq 2$ we denote $\mathbf{T}_E \circ \mathcal{H} : \mathbf{X} \to \mathbb{R}^+$ with the return time $\mathbf{T}_E$ defined in (6.9) by $\mathbf{T}_E$, too. We know from Lemma 6.4 that for E large,

$$0 < t_{\min} < \mathbf{T}_E < t_{\max}.$$

Moreover, being defined by composition of a smooth map with a Hölder map, $\mathbf{T}_E$ is Hölder.

To model the geodesic flow restricted to the set $\mathbf{b}_E \subset \boldsymbol{\Sigma}_E$ of bounded flow lines, we introduce the space

$$\mathbf{X}_E := \{(f,t) \mid f \in \mathbf{X}, 0 \leq t \leq \mathbf{T}_E(f)\}, \tag{6.34}$$

where we identify the points $(f, \mathbf{T}_E(f))$ with $(\sigma(f), 0)$. $\mathbf{X}_E$ will serve as a space modelling the set $\mathbf{b}_E$ of bounded geodesics.

Then the $\mathbf{T}_E$ *suspension flow* $\boldsymbol{\sigma}_E^t : \mathbf{X}_E \to \mathbf{X}_E$ is given by $\boldsymbol{\sigma}_E^t(f,r) := (f, r+t)$, up to the above identifications.

Now we are ready to describe the high-energy bounded orbits using symbolic dynamics. We know that the set $\mathbf{b}_E \subset \boldsymbol{\Sigma}_E$ consisting of bounded geodesic flow lines is of the form

$$\mathbf{b}_E = \left\{ x \in \boldsymbol{\Sigma}_E \mid \boldsymbol{\Phi}_E^t(x) \subset \mathbf{U}_E \; \forall t \in \mathbb{R} \right\}, \tag{6.35}$$

and $\mathbf{b}_E$ intersects the domain $\cup_{k=1}^{2(n-1)} C_k$ of the Poincaré map $\mathbf{P}$ in the set Λ analysed in Lemma 6.8.

Proposition 6.9 *For $n \geq 2$ and E large the map $\mathcal{H}_E : \mathbf{X}_E \to \mathbf{b}_E$ given by*

$$\mathcal{H}_E(f,t) := \boldsymbol{\Phi}_E^t(\mathcal{H}(f))$$

is a homeomorphism conjugating the suspension flow with the geodesic flow

$$\mathcal{H}_E \circ \boldsymbol{\sigma}_E^t = \boldsymbol{\Phi}_E^t \circ \mathcal{H}_E. \tag{6.36}$$

Thus, for $n = 2$, $\mathbf{b}_E$ consists of two closed geodesic flow lines, whereas for $n \geq 3$, locally $\mathbf{b}_E$ is homeomorphic to the product of a Cantor set and an interval. For $n = 1$, $\mathbf{b}_E = \emptyset$.

All bounded geodesics are hyperbolic and contain no focal points. $\mathbf{b}_E$ has measure zero w.r.t. the natural measure on $\boldsymbol{\Sigma}_E$.

Remark 6.10 *Hyperbolicity* of a geodesic flow line $\{\boldsymbol{\Phi}_E^t(x) \in \boldsymbol{\Sigma}_E \mid t \in \mathbb{R}\}$ as defined, e.g., in Def. 3.2.10 of [26] means the existence of a flow-invariant splitting of the tangent spaces along that flow line into the flow direction, an exponentially contracting and an exponentially expanding direction.

Proof. The map $\mathcal{H}_E$ is well-defined and bijective by definition (6.10) of the Poincaré map $\mathbf{P}$. It is a homeomorphism by continuity of the flow $\boldsymbol{\Phi}_E^t$, and by Lemma 6.8. (6.36) is immediate from (6.32).

For $n = 2$, the symbol space $\mathcal{S}$ of (6.15) consists of two symbols, and by definition (6.16) the transition matrix $\mathcal{M} = \begin{pmatrix} 1 & 0 \\ 0 & 1 \end{pmatrix}$, so that the space $\mathbf{X}$ of admissible sequences consists only of two elements.

For $n \geq 3$, the transition matrix $\mathcal{M}$ is *irreducible* and *aperiodic*, that is, there exists $m > 0$ for which all entries of $\mathcal{M}^m$ are strictly positive (choose any $m \geq 2$). Then being a subset of $\{f : \mathbb{Z} \to \mathcal{S}\}$, $\mathbf{X}$ is totally disconnected, and it is a (non-void) compact set without isolated points. Such topological spaces are homeomorphic to the Cantor set (see, *e.g.*, Franz, [13], Thm. 33.4).

As we remarked in the context of the construction of an invariant cone field, $\mathbf{b}_E = \emptyset$ for $n = 1$ and E large, since in that case the length of geodesic segments in $\mathbf{G}_E$ is bounded from above by $6R_{\mathrm{vir}}(E)$.

The existence of a strictly invariant cone field implies the hyperbolicity of the flow. The expanding subspace may be constructed as the intersection of all positive time-translates of cones.

The non-existence of focal points is immediate from Lemma 6.3.

The measure of the bounded orbits for E large may be estimated from above using the bound (6.21) for the width of the strips $W(k_{-m},\ldots,k_0)$ and summing over the admissible sequences $(k_{-m},\ldots,k_0)$. This bound goes to zero for $m\to\infty$. □

Now we analyse the set $b_E := b\cap\Sigma_E$ of bounded orbits of energy E in the original phase space P, using Prop. 6.9 and the covering construction of Prop. 3.1.

The first observation is that the lifted covering transformation $G^*:\boldsymbol{\Sigma}_E\to\boldsymbol{\Sigma}_E$ leaves the set of bounded geodesic flow lines invariant:

$$G^*(\mathbf{b}_E)=\mathbf{b}_E. \tag{6.37}$$

Moreover, the Poincaré surfaces C_k defined in (6.7), (6.8) are transformed according to the rule

$$G^*(C_k)=C_l \quad \text{with } |k-l|=n-1, \tag{6.38}$$

since $G:\mathbf{M}\to\mathbf{M}$ inverts the orientation of the geodesic segments $\mathbf{d}_k$ to which the C_k project (see Lemma 6.2).

By (6.37), (6.38) and the relation

$$\Lambda=\mathbf{b}_E\cap\left(\bigcup_{k=1}^{2(n-1)}C_k\right),$$

we have $G^*(\Lambda)=\Lambda$, and since $\mathcal{H}:\mathbf{X}\to\Lambda$ is a homeomorphism, we obtain a $\mathbb{Z}_2$-action

$$\mathcal{H}^{-1}\circ G^*\circ\mathcal{H}:\mathbf{X}\to\mathbf{X}$$

which we denote by G^*, too. By (6.38), for $f\in\mathbf{X}$ we have

$$\forall i\in\mathbb{Z}:\quad |(G^*f)(i)-f(i)|=n-1.$$

Now we define a space X_E modelling the set b_E of bounded orbits of energy E. Denoting an element $\{f,G^*f\}\in\mathbf{X}/\mathbb{Z}_2$ by $[f]$, let

$$X_E:=\{([f],t)\mid f\in\mathbf{X},0\le t\le T_E([f])\}, \tag{6.39}$$

where the reparametrized return time $T_E:\mathbf{X}/\mathbb{Z}_2\to\mathbb{R}^+$ is given by

$$T_E([f]):=s(\mathbf{T}_E(f),\mathcal{H}(f)) \tag{6.40}$$

with the time transformation $s:\mathbb{R}\times\boldsymbol{\Sigma}_E\to\mathbb{R}$ introduced in Prop. 3.1. Again, we identify points $([f],T_E([f]))$ with $([\sigma\circ f],0)$, and we define the T_E *suspension flow* $\sigma_E^t:X_E\to X_E$ by $\sigma_E^t(f,r):=(f,r+t)$, up to the above identifications.

Theorem 6.11 *For $n \geq 2$ and E large, the map $\mathcal{H}_E : X_E \to b_E$ given by*

$$\mathcal{H}_E([f], t) := \Phi_E^t(\pi_E(\boldsymbol{\mathcal{H}}(f)))$$

with π_E defined in Prop. 3.1 is a homeomorphism.

Furthermore, letting $g(f, t) := ([f], s(t, \boldsymbol{\mathcal{H}}(f))$, the diagram

$$\begin{array}{ccc} \mathbf{b}_E & \xleftarrow{\;\mathcal{H}_E\;} & \mathbf{X}_E \\ \big\downarrow{\scriptstyle \pi_E} & & \big\downarrow{\scriptstyle g} \\ b_E & \xleftarrow[\;\mathcal{H}_E\;]{} & X_E \end{array}$$

commutes. $\mathcal{H}_E$ conjugates the suspension flow and the flow Φ_E^t on b_E:

$$\mathcal{H}_E \circ \sigma_E^t = \Phi_E^t \circ \mathcal{H}_E.$$

Thus, for $n = 2$, b_E consists of one closed orbit, whereas for $n \geq 3$, locally b_E is homeomorphic to the product of a Cantor set and an interval. For $n = 1$, $\mathbf{b}_E = \emptyset$.

All bounded orbits are hyperbolic. b_E has measure zero w.r.t. the Liouville measure λ_E on Σ_E.

Proof. The theorem is an immediate consequence of Prop. 6.9 and our covering construction. □

Remarks 6.12

1. The measure of *all* positive energy bounded orbits is larger than zero in general. As an example, consider the case of a potential $V(\vec{q}) = -Z \exp(-\mu\, |\vec{q}\,|)/\, |\vec{q}\,|$, $\mu > 0$ of Yukawa form, see Chap. 11.

2. Clearly, the points on periodic orbits are in one-to-one correspondence with points in X_E with a periodic sequence of symbols.

   More interesting is the following observation: Let $([f], t) \in X_E$ be a point with a symbol sequence $f = (\ldots, a_{-1}, a_0, a_1 \ldots)$ which has a symmetry axis in the sense that for some $l \in \mathbb{Z}$

   $$a_{l+k} = a_{l-k} \qquad \text{for all } k \in \mathbb{Z}.$$

   Exactly in that case $\mathcal{H}_E([f], t)$ belongs to an orbit which eventually collides with a nucleus.

   The statement follows from a comparison of the fixed points of the joint action of time inversion and G^* on the energy shell $\boldsymbol{\Sigma}_E$ respectively on the symbol space.

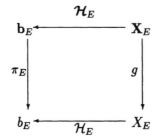

3. For E large there is a natural action of the braid group on n strings on the set $b_E/\sim$ of bounded orbits (with two points $x_1, x_2 \in b_E$ being equivalent if they belong to the same orbit). This action is nontrivial if $n \geq 3$.

   The *braid group on n strings* of a two-dimensional manifold M is by definition the fundamental group $\pi_1(M_n/\sim)$ with

$$M_n := \left\{(q_1, \dots, q_n) \in \prod_{l=1}^{n} M \,|q_i \neq q_j \text{ for } i \neq j\right\}.$$

   Two points $z, z' \in M_n$ are *equivalent* ($z \sim z'$) if there exists a permutation π such that $z'_l = z_{\pi(l)}$ for $l \in \{1, \dots, n\}$. See Birman [4] for more on braid groups.

   In our case M is the plane. Let

$$f \equiv (f_1, \dots, f_n) : I \to M_n$$

   be a path beginning at $f(0) = (\vec{s}_1, \dots, \vec{s}_n)$ and ending at $f(1) = (\vec{s}_{\pi(1)}, \dots, \vec{s}_{\pi(n)})$ for some permutation π. So f moves around the positions $\vec{s}_l$ of the nuclei and may be considered a representative of $[f] \in \pi_1(M_n/\sim, x)$ with base point $x := (\vec{s}_1, \dots, \vec{s}_n)/\sim \in M_n/\sim$.

   If we begin with a purely Coulombic potential of the form $V_0(\vec{q}) := -Z\sum_{l=1}^{n} 1/|\vec{q} - \vec{s}_l|$ and deform it by defining V_t, $t \in I$, to be the purely Coulombic potential

$$V_t(\vec{q}) := -\sum_{l=1}^{n} \frac{Z}{|\vec{q} - f_l(t)|},$$

   then $V_1 = V_0$ so that we performed a loop in the space of Coulombic potentials.

   For energy E we obtain for the Hamiltonians $\frac{1}{2}\vec{p}^2 + V_t(\vec{q})$ the bound states $b_E(t)$, $t \in I$, and the individual orbits vary continuously with t. Contractible loops in the space of Coulombic potentials do not change the orbits. So we obtain an action of the braid group on the set $b_E(0)/\sim$ of bounded orbits.

   We can modify this prescription for more general Coulombic potentials by transforming V into V_0 before and after applying f. For details in a similar case see [27].

4. In spite of asymptotic completeness, for $n \geq 2$ there is a measure zero set of orbits which come from spatial infinity and are trapped inside the interaction zone. Here we discuss the structure of this subset $s^- \setminus s \subset P$ of phase space; the set $s^+ \setminus s$ of orbits captured in negative time is related to $s^- \setminus s$ by time inversion.

   Fixing a large energy E, the asymptotic data, *i.e.* the incoming angular momentum and angle of these captured orbits, are given by the set

$$\mathcal{Z}_E^- := \{(L,\varphi) \in \mathbb{R}\times S^1 \mid \exists x \in (s^-\setminus s)\cap \Sigma_E \text{ with } L = L^-(x), \varphi = \varphi^-(x)\},$$

where the incoming angle $\varphi^-(x)$ is related to the incoming momentum $\vec{p}^-(x)$ by $p_1^-(x) = \sqrt{2E}\cos\varphi^-(x)$, $p_2^-(x) = \sqrt{2E}\sin\varphi^-(x)$.

We assume for simplicity the validity of eqs. (2.35) so that the asymptotic data vary smoothly with x.

Then for $n = 2$, $\mathcal{Z}_E^- \subset \mathbb{R} \times S^1$ is a smooth submanifold consisting of two components diffeomorphic to circles S^1 which wrap once around the cylinder. In other words, for any initial direction φ there are (at least) two impact parameters for which the corresponding orbit is captured by the interaction zone.

For $n \geq 3$ centres the picture is more complicated, and $\mathcal{Z}_E^-$ has the local form of a Cantor set times an interval.

The result for $n = 2$ is obtained by noticing that the set b_E of energy E bound states consists of a single closed unstable orbit. As a consequence of Thm. 7.2.8 of [1], its local stable manifold is an embedded two-dimensional submanifold of the energy shell Σ_E. Moreover (as can be seen using the above analysis of the Poincaré map), the stable manifold $(s^- \setminus s) \cap \Sigma_E$ can be globally embedded in Σ_E and extends to spatial infinity.

For $n \geq 3$ one may use the Stable Manifold Master Theorem (Thm. 7.2.8 of [1]) to analyse the structure of the captured orbits near the set b_E of bounded orbits.

In Figures 10.2, 10.3 we see the $\varphi \equiv 0$ slice of the cylinder $\mathbb{R} \times S^1$ of asymptotic data. There, points in $\mathcal{Z}_E^-$ correspond to data for which the time delay τ^- diverges.

7. Topological Entropy

Our next task will be to determine the topological entropy of the flow $\Phi_E^t = \Phi^t|_{\Sigma_E}$ on Σ_E. First we study the related question for the flow $\boldsymbol{\Phi}_E^t$ on $\boldsymbol{\Sigma}_E$.

Topological entropy is a quantity which, roughly speaking, measures the information loss per time unit about the state of the system. In the next chapter we shall relate that quantity to the dependence on T of the number of closed geodesics of minimal period $\leq T$.

The main technical tool of this chapter is Lemma 7.4 stating that the topological entropy of the geodesic flow is determined by its restriction to the set $\mathbf{b}_E$ of bounded geodesic flow lines. It is relatively easy to compute that topological entropy using symbolic dynamics.

First we formally introduce the notion of topological entropy, following the definition of Bowen. That definition is in a way more general than others in allowing for non-compact metric spaces (X, d) (see Walters [47]).

The topological entropy will then be invariant under uniformly continuous changes of the metric d. Therefore, for compact metrisable spaces X the topological entropy will only depend on the topology generated by the metric.

Definition 7.1 Let (X, d) be a metric space and $T : X \to X$ a uniformly continuous map. Then for $m \in \mathbb{N}$, $\varepsilon > 0$, a subset $F \subset X$ is said to (m, ε)-*span* a compact $K \subset X$ if

$$\forall x \in K\, \exists y \in F : d_m(x, y) \leq \varepsilon,$$

with the metric $d_m(x, y) := \max_{0 \leq i \leq m-1} d(T^i x, T^i y)$.

Let $r_m(\varepsilon, T, K)$ be the smallest cardinality of an (m, ε)-spanning set F of K,

$$h_r(\varepsilon, T, K) := \limsup_{m \to \infty} \frac{1}{m} \ln(r_m(\varepsilon, T, K)),$$

and $h_{\text{top}}(T, K) := \lim_{\varepsilon \to 0} h_r(\varepsilon, T, K)$.

Then the *topological entropy of* T is

$$h_{\text{top}}(T) := \sup\left\{h_{\text{top}}(T, K) \mid K \subset X \text{ compact}\right\}.$$

Note that $h_r(\varepsilon, T, K)$ is monotonically increasing in ε, so that its limit $h_{\text{top}}(T, K)$ exists (it can be ∞).

In estimating the topological entropy, the relations

$$h_{\rm top}(T^m, K) = m h_{\rm top}(T, K), \quad m \in \mathbb{N} \tag{7.1}$$

and

$$h_{\rm top}(T, T^m(K)) = h_{\rm top}(T, K), \quad m \in \mathbb{Z} \tag{7.2}$$

will turn out to be useful.

Equation (7.1) is contained in the proof of Thm. 7.10 of [47], whereas one can derive (7.2) by a suitable choice of (n, ε)-spanning sets for K and $T^m(K)$. Notice that in general $h_{\rm top}(T^{-1}, K)$ and $h_{\rm top}(T, K)$ are unrelated.

It will be technically helpful to introduce (m, ε)-*separated* sets:

Definition 7.2 Under the conditions of Definition 7.1 a subset $E \subset K$ is said to be (m, ε)-*separated* w.r.t. T if $x, y \in E$, $x \neq y$ implies $d_m(x, y) > \varepsilon$.

Let $s_m(\varepsilon, T, K)$ be the largest cardinality of any (m, ε)-separated set $E \subset K$ and

$$h_s(\varepsilon, T, K) := \limsup_{m \to \infty} \frac{1}{m} \ln(s_m(\varepsilon, T, K)).$$

The basic inequality

$$r_m(\varepsilon, T, K) \leq s_m(\varepsilon, T, K) \leq r_m(\varepsilon/2, T, K)$$

implies that $\lim_{\varepsilon \to 0} h_s(\varepsilon, T, K) = h_{\rm top}(T, K)$. So (m, ε)-separated sets are an alternative way to determine the topological entropy.

We are to estimate the topological entropy for $X := \Sigma_E$, $T := \Phi_E^1$, and for the metric on Σ_E induced by its Riemannian structure. T is uniformly continuous since the motion is asymptotically free (see below).

In our case we have

$$h_{\rm top}(\Phi_E^{-1}) = h_{\rm top}(\Phi_E^{+1}),$$

although in general $h_{\rm top}(\Phi_E^{-1}, K) \neq h_{\rm top}(\Phi_E^{+1}, K)$, since we have a symmetry w.r.t. time reversal.

Lemma 7.4 below will show the importance of the set $\mathbf{b}_E$ of bounded geodesic flow lines.

Definition 7.3 The *non-wandering set* $\Omega(\Phi)$ of a continuous flow $\Phi^t : X \to X$ is given by

$$\Omega(\Phi) := \left\{ x \in X \;\middle|\; \forall \text{ neighb. } U \ni x, \forall T > 0 : U \cap \left(\bigcup_{t \geq T} \Phi^t(U) \right) \neq \emptyset \right\}.$$

For non-compact spaces X the topological entropy is in general larger than the topological entropy of the restriction to the non-wandering set. Nevertheless, in our case we have

Lemma 7.4 *For $E > V_{\max}$, $\Omega(\boldsymbol{\Phi}_E) \subset \mathbf{b}_E$. Furthermore, for E sufficiently large, $\Omega(\boldsymbol{\Phi}_E) = \mathbf{b}_E$ and the topological entropy of the geodesic flow is determined by its bounded geodesic flow lines, that is,*

$$h_{\text{top}}(\boldsymbol{\Phi}_E^1) = h_{\text{top}}(\boldsymbol{\Phi}_E^1|_{\mathbf{b}_E}). \tag{7.3}$$

Proof. It is easy to see that $\Omega(\boldsymbol{\Phi}_E) \subset \mathbf{b}_E$ for $E > V_{\max}$ since we know by (6.35) that for $x \notin \mathbf{b}_E$ there exists a time $t \in \mathbb{R}$ with $\boldsymbol{\Phi}_E^t(x) \notin \mathbf{U}_E$. By the virial inequality all points outside $\mathbf{U}_E$ are wandering. Then x is wandering, too, since $\boldsymbol{\Phi}_E^t(x)$ is wandering which is a property shared by points on an orbit.

For E sufficiently large, we use symbolic dynamics to see that the set of periodic orbits (which are non-wandering) is dense in the set $\mathbf{b}_E$ of bounded orbits. Since $\Omega(\boldsymbol{\Phi}_E)$ is closed, we get $\Omega(\boldsymbol{\Phi}_E) = \mathbf{b}_E$.

To prove (7.3), we show that

$$h_{\text{top}}(\boldsymbol{\Phi}_E^1, K) = 0 \quad \text{for } K \subset \boldsymbol{\Sigma}_E \text{ compact and } K \cap \mathbf{b}_E^+ = \emptyset, \tag{7.4}$$

with

$$\mathbf{b}_E^+ := \left\{x \in \boldsymbol{\Sigma}_E \mid \exists t_0 \in \mathbb{R}\, \forall t \geq t_0 : \boldsymbol{\Phi}_E^t(x) \in \mathbf{U}_E\right\} \tag{7.5}$$

denoting the set of orbits captured in the interaction zone (and thus belonging to the stable manifolds of the bounded orbits).

It is clear that for every $x \in K$ there exists a time after which $\boldsymbol{\Phi}_E^t(x) \notin \mathbf{U}_E$. Moreover, for any radius $R > R_{\text{vir}}(E)$, there exists a time $t_0 = t_0(R, x)$ such that for all $t \geq t_0$, the radial distance $\left|\boldsymbol{\eta}_E(\boldsymbol{\Phi}_E^t(x))\right| \geq R$, by part 1 of Corollary 2.8. Since K is compact, we can find a common $t_0 = t_0(R, K)$ valid for all $x \in K$. In other words, after time t_0 the set $\boldsymbol{\Phi}_E^t(K)$ has a large distance to the interaction zone and moves to spatial infinity. There the motion is nearly free since the metric $\mathbf{g}_E$ is nearly flat: $|K_E(q)| < \frac{C_1}{E} R^{-2-\varepsilon}$ by (3.18).

We apply the identity $h_{\text{top}}(\boldsymbol{\Phi}_E^1, \boldsymbol{\Phi}_E^m(K)) = h_{\text{top}}(\boldsymbol{\Phi}_E^1, K)$ (see (7.2)) with $m \geq t_0(R, K)$. Moreover, $h_{\text{top}}(\boldsymbol{\Phi}_E^1, \boldsymbol{\Phi}_E^m(K)) = \frac{1}{l} h_{\text{top}}(\boldsymbol{\Phi}_E^l, K)$ for $l \geq 1$ by (7.1). Then as a consequence of the proof of Thm. 7.15 in [47], we can bound $h_{\text{top}}(\boldsymbol{\Phi}_E^1, \boldsymbol{\Phi}_E^m(K))$ from above by

$$h_{\text{top}}(\boldsymbol{\Phi}_E^1, \boldsymbol{\Phi}_E^m(K)) \leq \frac{2}{l} \ln\left(\sup\left\{\|D\boldsymbol{\Phi}_E^l(x)\| \mid x \in \boldsymbol{\Sigma}_E, |\boldsymbol{\eta}_E(x)| \geq R\right\}\right), \tag{7.6}$$

where $D\boldsymbol{\Phi}_E^l(x) : T_x\boldsymbol{\Sigma}_E \to T_y\boldsymbol{\Sigma}_E$, $y := \boldsymbol{\Phi}_E^l(x)$, is the linearization of the time l flow at x. The operator norm $\|\cdot\|$ is given by

$$\|D\boldsymbol{\Phi}_E^l(x)\| := \max\left\{\frac{\|D\boldsymbol{\Phi}_E^l(x)(v)\|}{\|v\|} \,\middle|\, v \in T_x\boldsymbol{\Sigma}_E, v \neq 0\right\},$$

$\|v\|$ denoting the length of v in the metric on $\boldsymbol{\Sigma}_E$ induced by $\mathbf{g}_E$.

It suffices to estimate the operator norm for vectors $v \in T_x\boldsymbol{\Sigma}_E$ perpendicular to the direction of the flow, by flow-invariance of the Hamiltonian vector field generating $\boldsymbol{\Phi}_E^t$, and by constancy of its norm: $\|D\boldsymbol{\Phi}_E^l(x)(w)\| = \|w\|$ for w tangent to the flow (Lemma 3.1.15 of [26]).

Therefore, we are to estimate the norm of the monodromy matrix $U(l)$ for the Jacobi equation

$$\frac{d}{dt}U(t) = \begin{pmatrix} 0 & 1 \\ -K(t) & 0 \end{pmatrix} U(t), \qquad U(0) := \mathbf{1} \tag{7.7}$$

with $K(t) := K_E(\boldsymbol{\eta}_E(\boldsymbol{\Phi}_E^t(x)))$ denoting the Gaussian curvature along the geodesic.

To obtain a good estimate which exploits the curvature estimate (3.18), we multiply (7.7) from the left with the matrix

$$B := \begin{pmatrix} 1 & 0 \\ 0 & b \end{pmatrix}, \qquad \text{with } b := \sqrt{E/C_1}R.$$

Then $V(t) = BU(t)$ solves

$$\frac{d}{dt}V(t) = \begin{pmatrix} 0 & b^{-1} \\ -bK(t) & 0 \end{pmatrix} V(t), \qquad V(0) := B.$$

For R large, one has $b > 1$. Gronwall's inequality and the curvature estimate (3.18) lead to the estimate

$$\|U(t)\| \leq \|B^{-1}\| \|V(t)\| = \|V(t)\| \leq b \exp(t/b).$$

Thus by (7.6)

$$h_{\text{top}}(\boldsymbol{\Phi}_E^1, K) = \frac{1}{l} h_{\text{top}}(\boldsymbol{\Phi}_E^l, \boldsymbol{\Phi}_E^m(K)) \leq \frac{2}{l}(\ln b + l/b) < 3/b \text{ for } l \text{ large},$$

which shows (7.4) since b is proportional to R and may be chosen arbitrarily large by enlarging m.

So to calculate the topological entropy $h_{\text{top}}(\boldsymbol{\Phi}_E^1)$ of the geodesic flow, we only need to take into account compacts $K \subset \boldsymbol{\Sigma}_E$ having nonzero intersection with the stable manifold $\mathbf{b}_E^+$ defined in (7.5).

We will use the inequality

$$h_{\text{top}}(\boldsymbol{\Phi}_E^1, K) \leq \max_{1 \leq i \leq m} h_{\text{top}}(\boldsymbol{\Phi}_E^1, K_i) \tag{7.8}$$

valid for the compacts K, K_i with $K \subset \bigcup_{i=1}^m K_i$ (Thm. 7.5 of [47]).

(7.8), (7.2) and the known properties of the flow $\boldsymbol{\Phi}_E^t$ imply for each compact K the existence of a $t_0 > 0$ such that

$$h_{\text{top}}(\boldsymbol{\Phi}_E^1, K) \leq h_{\text{top}}(\boldsymbol{\Phi}_E^1, \boldsymbol{\Phi}_E^m(K) \cap \mathbf{U}_E) \text{ for all } m \geq t_0 \tag{7.9}$$

since after time t_0 no point $x \in K$ enters $\mathbf{U}_E$, and since the entropy

$$h_{\text{top}}\left(\boldsymbol{\Phi}_E^1, \boldsymbol{\Phi}_E^m(K) \cap (\boldsymbol{\Sigma}_E \setminus \text{Int}(\mathbf{U}_E))\right) = 0$$

by (7.4).

By (7.9) we need only consider compacts $K \subset \mathbf{U}_E$, so that it is sufficient to show that

$$h_{\text{top}}(\boldsymbol{\Phi}_E^1, \mathbf{b}_E) = h_{\text{top}}(\boldsymbol{\Phi}_E^1, \mathbf{U}_E). \tag{7.10}$$

Clearly, $h_{\text{top}}(\boldsymbol{\Phi}_E^1, \mathbf{b}_E) \leq h_{\text{top}}(\boldsymbol{\Phi}_E^1, \mathbf{U}_E)$ since $\mathbf{b}_E \subset \mathbf{U}_E$. Therefore we are done in the case $h_{\text{top}}(\boldsymbol{\Phi}_E^1, \mathbf{U}_E) = 0$ so that we assume from now on that

$$h_{\text{top}}(\boldsymbol{\Phi}_E^1, \mathbf{U}_E) > 0 \tag{7.11}$$

which implies $h_s(\varepsilon, \boldsymbol{\Phi}_E^1, \mathbf{U}_E) > 0$ for small $\varepsilon > 0$.

Let $K_1 := \mathbf{U}_E$ and $K_2 := \boldsymbol{\Phi}_E^1(\mathbf{U}_E) \setminus \text{Int}(\mathbf{U}_E)$ (so K_1 and K_2 are compact).

Let $E_m \subset K_1$ be a (m, ε)-separating set for K_1 of maximal cardinality (introduced in Def. 7.2), that is, $\#E_m = s_m(\varepsilon, \boldsymbol{\Phi}_E^1, K_1)$. We partition E_m into m disjoint subsets by writing

$$E_m = \bigcup_{r=1}^{m-1} E_m^r \, \cup R_m$$

where $e \in E_m^r$ if and only if

$$\boldsymbol{\Phi}_E^{r-1}(e) \in K_1 \text{ and } \boldsymbol{\Phi}_E^r(e) \notin K_1.$$

We are to show that the number $\tilde{s}_m := \sum_{r=1}^{m-1} \#E_m^r$ of elements of E_m which leave K_1 within $m-1$ iterations is small compared to the total number $s_m = \#E_m$ of elements.

Clearly, the subset $\boldsymbol{\Phi}_E^r(E_m^r) \subset K_2$ is a $(m-r, \varepsilon)$-separating set for K_2.

We know by (7.4) that $h_{\text{top}}(\boldsymbol{\Phi}_E^1, K_2) = 0$ since $K_2 \cap \mathbf{b}_E^+ = 0$, as there are no points reentering $\mathbf{U}_E$ after leaving it. The general inequality

$$h_s(\varepsilon_1, T, K) \geq h_s(\varepsilon_2, T, K), \text{ valid for } \varepsilon_1 < \varepsilon_2,$$

implies $h_s(\varepsilon, \boldsymbol{\Phi}_E^1, K_2) = 0$ for $\varepsilon > 0$.

Thus for any $\delta > 0$ there exists a $m_0(\delta) \in \mathbb{N}$ with

$$\frac{1}{m-1} \ln(s_m(\varepsilon, \boldsymbol{\Phi}_E^1, K_2)) < \delta \text{ for } m \geq m_0(\delta).$$

Let $m > m_0(\delta)$. Then

$$\begin{aligned}
\frac{1}{m-1} \ln(\tilde{s}_m) &= \frac{1}{m-1} \ln\left(\sum_{r=1}^{m-1} \#E_m^r\right) \\
&\leq \frac{1}{m-1} \ln\left(\sum_{r=1}^{m-1} s_{m-r}(\varepsilon, \boldsymbol{\Phi}_E^1, K_2)\right) \\
&\leq \frac{1}{m-1} \ln((m-1) \cdot s_{m-1}(\varepsilon, \boldsymbol{\Phi}_E^1, K_2)) \\
&< \delta + \frac{\ln(m-1)}{m-1}
\end{aligned}$$

which shows, using (7.11), that $\lim_{m\to\infty} \ln(\tilde{s}_m/s_m) = 0$.

In other words, the subset $R_m \subset E_m$ of points staying in K_1 for at least $m-1$ iterations is of density one.

We shall now use this result to finish the proof of Lemma 7.4 by a measure theoretic argument. Let $\sigma_m := (s_m(\varepsilon, \boldsymbol{\Phi}_E^1, K_1))^{-1} \sum_{x\in E_m} \delta_x$ be the atomic probability measure concentrated uniformly on the points of E_m, and define the probability measure μ_m on $\boldsymbol{\Sigma}_E$ by

$$\mu_m := \frac{1}{m} \sum_{r=1}^{m-1} \sigma_m \circ \boldsymbol{\Phi}_E^{-r}.$$

Then by the above result, $\lim_{m\to\infty} \mu_m(K_1) = 1$.

By compactness of K_1, the space $M(K_1)$ of probability measures on K_1 is compact in the weak*-topology, see [47]. Therefore we can find a subsequence $\{m_j\}$ of natural numbers such that

$$\lim_{j\to\infty} \frac{1}{m_j} \ln(s_{m_j}(\varepsilon, \boldsymbol{\Phi}_E^1, K_1)) = h_s(\varepsilon, \boldsymbol{\Phi}_E^1, K_1)$$

and at the same time μ_{m_j} converges weakly to a measure $\mu \in M(K_1)$.

We know from Thm. 6.9 of [47] (whose proof also works in our slightly different situation) that $\boldsymbol{\Phi}_E^1$ leaves μ invariant.

From Thm. 6.15 of [47] (or rather its proof) we know that transformation-invariant probability measures are concentrated on the non-wandering set of the transformation. In our case the non-wandering set of $\boldsymbol{\Phi}_E^1$ is contained in the set $\mathbf{b}_E$ of bounded orbits. Therefore $\mu(\mathbf{b}_E) = 1$.

Denoting by $h_\mu(\boldsymbol{\Phi}_E^1, \mathbf{b}_E)$ the measure-theoretic entropy of $\boldsymbol{\Phi}_E^1$, we have

$$h_\mu(\boldsymbol{\Phi}_E^1, \mathbf{b}_E) \geq h_s(\varepsilon, \boldsymbol{\Phi}_E^1, K_1) \tag{7.12}$$

precisely by the same proof as the one of Thm. 8.6 of [47]. Taking the limit $\varepsilon \to 0$, we have

$$h_\mu(\boldsymbol{\Phi}_E^1, \mathbf{b}_E) \geq h_{\text{top}}(\boldsymbol{\Phi}_E^1, K_1). \tag{7.13}$$

It is well-known (Thm 8.6 of [47]) that on compact spaces the measure-theoretic entropy is smaller or equal than the topological entropy. So we have

$$h_{\text{top}}(\boldsymbol{\Phi}_E^1, \mathbf{b}_E) \geq h_\mu(\boldsymbol{\Phi}_E^1, \mathbf{b}_E). \tag{7.14}$$

(7.13) and (7.14) imply (7.10). By Remark 10, p. 169 of [47] we know that $h_{\text{top}}(\boldsymbol{\Phi}_E^1, \mathbf{b}_E) = h_{\text{top}}(\boldsymbol{\Phi}_E^1|_{\mathbf{b}_E})$, which implies (7.3). □

Having finally Lemma 7.4 at our disposal, we are now in a situation to obtain the basic estimates on the topological entropy using symbolic dynamics.

Proposition 7.5 *For E large, the topological entropy $h_{\text{top}}(\boldsymbol{\Phi}_E^1)$ is bounded by*

$$\frac{\ln|2n-3|}{t_{\max}} \leq h_{\text{top}}(\boldsymbol{\Phi}_E^1) \leq \frac{\ln|2n-3|}{t_{\min}}. \tag{7.15}$$

Proof. We know from (7.3) that $h_{\text{top}}(\boldsymbol{\Phi}_E^1) = h_{\text{top}}(\boldsymbol{\Phi}_E^1|\mathbf{b}_E)$.

$$h_{\text{top}}(\boldsymbol{\Phi}_E^1|\mathbf{b}_E) = h_{\text{top}}(\sigma_{\mathbf{T}_E}^1)$$

since the geodesic flow on $\mathbf{b}_E$ is conjugate to the suspension flow $\sigma_{\mathbf{T}_E}^1$ on $\mathbf{X}_E$ w.r.t. to the homeomorphism $\mathcal{H}_E : \mathbf{X}_E \to \mathbf{b}_E$, and since for compact spaces topological entropy is an invariant under homeomorphisms (Thm. 7.2 of [47]).

For $n = 1$, the topological entropy is zero for E large, since $\mathbf{b}_E = \emptyset$, see Proposition 6.9.

For $n = 2$, we know from Proposition 6.9 that $\mathbf{b}_E$ consists of only two closed orbits. Therefore, by Thm. 7.14 of [47], the topological entropy is zero, too.

For $n \geq 3$, we know that the transition matrix $\mathcal{M}$ is irreducible. In such a case, by Thm 8.10 of [47], there exists a unique probability measure ν of maximal entropy $h_\nu(\sigma) = h_{\text{top}}(\sigma) = \ln(2n-3)$ for the shift $\sigma : \mathbf{X} \to \mathbf{X}$, called the *Parry measure.* Here $2n - 3$ is the largest positive eigenvalue of $\mathcal{M}$.

The idea is now to estimate the topological entropy of the time one flow $\sigma_{\mathbf{T}_E}^1$ on $\mathbf{X}_E$ by applying the formula

$$h_{\mu_L}(\sigma_{\mathbf{T}_E}^1) = \left(\int_{\mathbf{X}} \mathbf{T}_E(x)d\mu\right)^{-1} h_\mu(\sigma), \tag{7.16}$$

with μ_L being the normalized measure on $\mathbf{X}_E$ derived from any ergodic measure μ. (7.16) holds true by Theorem 2.1 of Chap. 3 in Sinai [40].

Taking for μ the Parry measure ν, and bounding $\int_{\mathbf{X}} \mathbf{T}_E(x)d\mu$ from above by $t_{\max}$, we obtain the l.h.s. of (7.15), observing that the measure-theoretic entropy is always smaller than the topological entropy.

The r.h.s. of (7.15) is derived by noticing that there exists a measure $\mu(\mathbf{T}_E)$ of maximal entropy on $\mathbf{X}_E$, since $\sigma_{\mathbf{T}_E}^1$ is expansive (Chap. 8.3 of [47]). This measure is unique since $\mathbf{T}_E$ is Hölder. Thus by Thm. 8.7 (iii) of [47] it is ergodic and we can apply (7.16), bounding $\int_{\mathbf{X}} \mathbf{T}_E(x)d\mu$ from below by $t_{\min}$. □

Now we are to determine the topological entropy $h_{\text{top}}(\Phi_E^1)$ of the original motion.

Since the energy shell Σ_E is not compact, the topological entropy is not an invariant of the metric distance. On the covering space $\boldsymbol{\Sigma}_E$ we had a natural metric distance derived from the Jacobi metric $\mathbf{g}_E$ on $\mathbf{M}$. Such a 'natural' metric does not exist on Σ_E. On the other hand, outside the 'interaction zone' near the nuclei, the metric distance dist (see (2.18)) is a natural choice.

The theorem below is valid for *any* metric on Σ_E which makes the flow Φ_E^t uniformly continuous and which is uniformly equivalent to dist near spatial infinity.

Theorem 7.6 *There exist $C_5 > C_6 > 0$ such that for E large, the topological entropy $h_{\text{top}}(\Phi_E^1)$ is bounded by*

$$C_6 \ln|2n-3|\sqrt{E} \leq h_{\text{top}}(\Phi_E^1) = h_{\text{top}}(\Phi_E^1|_{b_E}) \leq C_5 \ln|2n-3|\sqrt{E}. \tag{7.17}$$

Proof. The proof of the equality $h_{top}(\Phi_E^1) = h_{top}(\Phi_E^1|_{b_E})$ is the same as the proof of Lemma 7.4, except for the fact that we have to estimate the 4×4 matrix $U(t)$ with

$$\frac{d}{dt}U(t) = \begin{pmatrix} 0 & \mathbf{1} \\ -\partial_i\partial_j V(\vec{q}(t)) & 0 \end{pmatrix} U(t), \qquad U(0) := \mathbf{1} \tag{7.18}$$

instead of (6.14).

To estimate $h_{top}(\Phi_E^1|_{b_E})$, we use the equality $h_{top}(\Phi_E^1|_{b_E}) = h_{top}(\sigma_{T_E}^1)$ with the topological entropy of the time one map of the suspension flow. We proceed as in the proof of Prop. 7.5 and show (7.17) with the help of the kinematical inequality

$$\frac{k_1}{\sqrt{E}} \le T_E \le \frac{k_2}{\sqrt{E}}$$

valid for suitable $k_2 > k_1 > 0$ and E large. □

Remark 7.7 In principle we can determine the constants occurring in Theorem 7.6 with arbitrary precision (that is, for $\varepsilon > 0$ arbitrarily small, we may choose $C_5 > C_6$ so that $C_5 = (1+\varepsilon)C_6$). However, the price to be paid is then that the energy E must be larger than a threshold which goes to infinity as $\varepsilon \searrow 0$.

To obtain such estimates, we use a simple model for the high energy bounded orbits. We approximate an arbitrary bounded trajectory in the configuration plane by a path consisting of straight line segments which connect the positions $\vec{s}_{a_{l-1}}, \vec{s}_{a_l}$ of the nuclei.

The finite polygons are described by sequences $(a_0, \ldots, a_r)$ with $a_l \in \{1, \ldots, n\}$, the only restriction being that consecutive symbols are different.

This model is a good approximation of the real motion since a true trajectory passes the nuclei with a minimal distance which is of order $\mathcal{O}(E^{-1})$ (see Proposition 5.1).

The time between two such near-collisions is given by

$$(2E)^{-1/2} \cdot \left|\vec{s}_{a_{l-1}} - \vec{s}_{a_l}\right| \cdot (1 + \mathcal{O}(E^{-1})).$$

The topological entropy of the time-one shift of our model is given by

$$h_{top}^M := \lim_{L\to\infty} \frac{N(L)}{\sqrt{2E}L}$$

where

$$N(L) := \#\{(a_0, \ldots, a_r) \mid r \in \mathbb{N}, a_l \in \{1, \ldots, n\}, a_l \neq a_{l-1}, \sum_{l=1}^{r} \left|\vec{s}_{a_{l-1}} - \vec{s}_{a_l}\right| < L\}$$

is the number of polygons whose total length does not exceed L.

Then for E large, we have

$$h_{top}(\Phi_E^1) = h_{top}^M \cdot (1 + \mathcal{O}(E^{-1})).$$

8. The Distribution of the Closed Orbits

Our general aim is to analyse the typical behaviour of high-energy orbits. But the typical high-energy orbits are scattering states, since the bounded orbits are of measure zero by Proposition 6.9 and since the non-bounded non-scattering orbits are of measure zero by asymptotic completeness (Corollary 2.8). On the other hand, the bounded orbits *are* important since they influence the structure of those scattering states which have a long time delay. Lemma 7.4 gives a first example of a global observable (the topological entropy) which is determined entirely by the bounded orbits.

In this chapter we will continue the analysis of the bounded orbits.

One immediate consequence of the symbolic representation established in Proposition 6.9 is the fact that for $n \geq 3$ there are uncountably many bounded orbits of a given energy, whereas the subset of *closed* orbits is countable.

The easiest way of counting is in ascending order of the minimal periods. For Anosov flows and, more generally, Axiom A flows (that is, roughly speaking, flows on compact spaces whose non-wandering set carries a hyperbolic structure and is the closure of the periodic orbits) there is a direct relation between the growth rate of the number $N(T)$ of closed orbits with minimal period $\leq T$, and the topological entropy h_{top}, namely

$$h_{\mathrm{top}} = \lim_{T \to \infty} \frac{1}{T} \ln(N(T)). \tag{8.1}$$

Clearly, it is more subtle to derive the converse form of (8.1), namely to find an asymptotic formula for $N(T)$ in terms of the topological entropy.

If all periods are integral multiples of some 'minimal time' ΔS, then $N(T)$ is a step function of step width ΔS. In such a situation, we cannot even expect the existence of a smooth function of T to which $N(T)$ is asymptotic.

In [37], Parry and Pollicott derived a deep result for Axiom A flows which roughly speaking says that if no such ΔS exists, then

$$N(T) \sim \exp(h_{\mathrm{top}}T)/h_{\mathrm{top}}T,$$

i.e., $N(T) = (\exp(h_{\mathrm{top}}T)/h_{\mathrm{top}}T) \cdot (1 + o(1))$ as $T \to \infty$. Since their result readily generalizes to our situation of a non-compact energy shell, our basic task will be to show the non-existence of a minimal time ΔS.

Remark 8.1 Before returning to this somewhat subtle point, we analyse the number of closed orbits for iterates of the Poincaré map $\mathbf{P}_\Lambda$. This question is simple to answer (since maps are simpler than flows), and we use it as a warm-up.

By Lemma 6.8, $\mathbf{P}_\Lambda$ is conjugate to the shift σ on $\mathbf{X}$. Thus, instead of counting the number $N(k)$ of fixed points of $\mathbf{P}_\Lambda^k$, we may count the number of fixed points of σ^k, which is given by $N(k) = \mathrm{tr}(\mathcal{M}^k)$, with $\mathcal{M}$ defined in (6.16). To calculate the trace, we write the transition matrix $\mathcal{M}$ in the form $\mathcal{M} = F - J$, where $F_{ik} = 1$ for all $i, k \in \{1, \ldots, 2(n-1)\}$. Then J is a permutation matrix of order two ($J^2 = 1$) which implies the identities $JF = FJ = F$. Therefore,

$$\mathcal{M}^k = (-1)^k (J^k - 1) + (F - 1)^k$$

and

$$\begin{aligned} N(k) &= (2n-3)^k + n - 1 + (-1)^k (n-2) \\ &= \exp(h_{\mathrm{top}} \cdot k) + n - 1 + (-1)^k (n-2), \end{aligned}$$

with $h_{\mathrm{top}} \equiv h_{\mathrm{top}}(\sigma) = \ln(2n-3)$ as shown in the proof of Proposition 7.5.

So for $n = 2$; $N(k) = 2$ independent of k.

From now on we consider the cases $n \geq 3$.

The number $N_{\mathrm{min}}(k)$ of fixed points of *minimal* period k is of the same order:

$$N_{\mathrm{min}}(k) = N(k) + \mathcal{O}(N([k/2])),$$

since the minimal period of a fixed point of σ^k divides k.

The number of *orbits* of minimal period k equals $N_{\mathrm{min}}(k)/k$.

Therefore, the number $N(T)$ of orbits whose minimal period is strictly smaller than T equals

$$\begin{aligned} N(T) &= \sum_{k=1}^{T-1} \frac{N_{\mathrm{min}}(k)}{k} = \sum_{k=1}^{T-1} \frac{\exp(h_{\mathrm{top}} k)}{k} + \mathcal{O}\left(\exp(h_{\mathrm{top}} k/2)\right) \\ &= \frac{\exp(h_{\mathrm{top}} T)}{(\exp(h_{\mathrm{top}}) - 1) T} (1 + \mathcal{O}(1/T)). \end{aligned}$$

We return to the case of flows and start by considering the geodesic flow Φ_E^t, restricted to the set $\mathbf{b}_E$ of bounded orbits.

To apply the work [37] of Parry and Pollicott and the work [6] of Bowen on which [37] is based, we show that the stable and unstable manifolds of the points in $\mathbf{b}_E$ form a 'nonintegrable pair' similar to the case of Anosov flows analysed in [2].

We start with some definitions. For $x \in \mathbf{b}_E$

$$W^{s/u}(x) := \left\{ y \in \mathbf{b}_E \mid \lim_{t \to \pm\infty} d(\Phi_E^t(x), \Phi_E^t(y)) = 0 \right\} \tag{8.2}$$

denotes the stable resp. unstable set of x *within* $\mathbf{b}_E$, whereas

$$W_\varepsilon^{s/u}(x) := \left\{ y \in W^{s/u}(x) \mid \forall t > 0 \; d(\Phi_E^{\pm t}(x), \Phi_E^{\pm t}(y)) < \varepsilon \right\}.$$

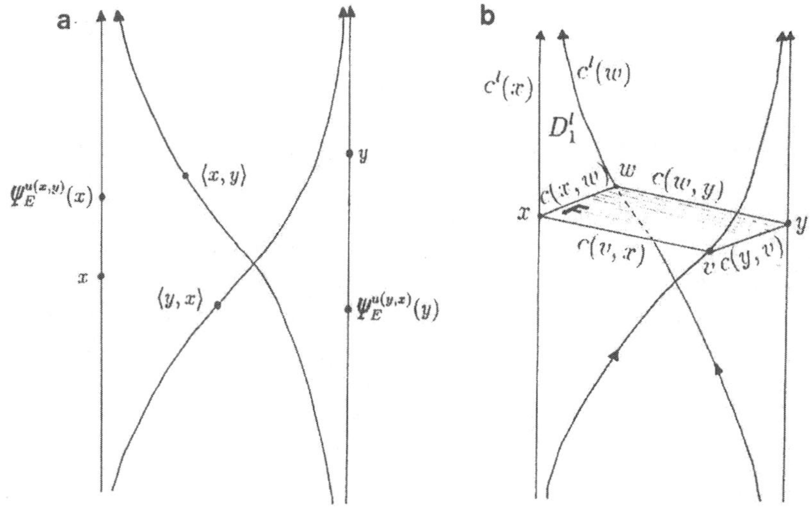

Fig. 8.1. a) Heteroclinic orbits connecting flow lines through x and y. b) The region F in the domain C_l of the Poincaré map.

The Hölder conjugacy of the flow $\boldsymbol{\Phi}_E^t$ and the suspension flow $\boldsymbol{\sigma}_E^t$ on $\mathbf{X}_E$ implies the existence of 'canonical coordinates' in the sense of [6]:

There are $\delta, \gamma > 0$ (possibly depending on the energy E) for which the following holds true:

If $x, y \in \mathbf{b}_E$ and $d(x,y) \leq \delta$, then there is a unique $u = u(x,y) \in \mathbb{R}$ with $|u| \leq \gamma$ so that

$$W_\gamma^s(\boldsymbol{\Phi}_E^u(x)) \cap W_\gamma^u(y) \tag{8.3}$$

consists of exactly one point which is then denoted by the ordered pair $\langle x, y\rangle$ (see Fig. 8.1 a)). Moreover, the maps u and $\langle \cdot, \cdot \rangle$ are continuous.

We are interested in the *desynchronization time*

$$\Delta T(x,y) := u(x,y) + u(y,x) \text{ on } \{(x,y) \in \mathbf{b}_E \times \mathbf{b}_E \mid d(x,y) \leq \delta\},$$

which measures the lack of integrability of the stable and unstable foliations.

Now we consider for simplicity points $x, y \in \mathbf{b}_E$ which are contained in the same component, say C_l, of the domain of the Poincaré map, see (6.7), (6.8).

We denote by $f_x := \mathcal{H}^{-1}(x)$ and $f_y := \mathcal{H}^{-1}(y)$ the symbol sequences of x and y, respectively. By restriction, we obtain the half-infinite sequences $f_x^\pm$ and $f_y^\pm$ which by (6.29) and (6.30) fix the curves

$$V(f_x^+), W(f_y^-), V(f_y^+), W(f_x^-) \subset C_l. \tag{8.4}$$

The (oriented) region in C_l enclosed by subintervals of these curves is denoted by F, with the orientation induced by the succession of curves in (8.4), see Fig. 8.1 b). The region F is thus diffeomorphic to a (possibly degenerate) rectangle.

The desynchronization time $\Delta T(x,y)$ has a direct geometric interpretation:

Lemma 8.2 *In the situation described above,*

$$\Delta T(x,y) = -\int_F \omega_E, \tag{8.5}$$

where ω_E denotes the symplectic two-form on $(T\mathbf{M}, \mathbf{g}_E)$.

Proof. We use the coordinatization of the rectangle C_l by $(l, \cos\alpha)$ introduced in Chap. 6, and write $x = (l_x, \cos\alpha_x)$, $y = (l_y, \cos\alpha_y)$. Similarly, we denote the (unique) point $W(f_x^-) \cap V(f_y^+)$ by $v = (l_v, \cos\alpha_v)$ and $W(f_y^-) \cap V(f_x^+)$ by $w = (l_w, \cos\alpha_w)$. We calculate the desynchronization time using the restriction $\mathbf{P}_\Lambda = \mathbf{P}|_\Lambda$ of the Poincaré map to the invariant set $\Lambda = \Lambda^+ \cap \Lambda^-$ introduced in (6.31).

$$u(x,y) = \sum_{k=0}^{\infty}\left(\mathbf{T}_E(\mathbf{P}_\Lambda^k(x)) - \mathbf{T}_E(\mathbf{P}_\Lambda^k(w))\right) - \sum_{k=-1}^{-\infty}\left(\mathbf{T}_E(\mathbf{P}_\Lambda^k(w)) - \mathbf{T}_E(\mathbf{P}_\Lambda^k(y))\right) \tag{8.6}$$

and

$$u(y,x) = \sum_{k=0}^{\infty}\left(\mathbf{T}_E(\mathbf{P}_\Lambda^k(y)) - \mathbf{T}_E(\mathbf{P}_\Lambda^k(v))\right) - \sum_{k=-1}^{-\infty}\left(\mathbf{T}_E(\mathbf{P}_\Lambda^k(v)) - \mathbf{T}_E(\mathbf{P}_\Lambda^k(x))\right) \tag{8.7}$$

with Poincaré map $\mathbf{P}_\Lambda$ and return time $\mathbf{T}_E$ defined in (6.9). The sums in (8.6) and (8.6) are converging exponentially fast since the return time $\mathbf{T}_E$ is smooth and since points on (un)stable manifolds converge exponentially fast.

Now we rewrite these sums in the form of integrals over the symplectic one-form θ which locally equals $\theta = p_1 dq^1 + p_2 dq^2$ (in cotangent variables) whose exterior derivative is $-d\theta = \omega_E$.

For a point $z \in \Lambda^+$ on a stable manifold and $l \geq 0$ the time $T^l(z) := \sum_{k=0}^{l} \mathbf{T}_E(\mathbf{P}^k(z))$ is well-defined. Considering the first term in (8.6), we get

$$\begin{aligned}\sum_{k=0}^{\infty}\left(\mathbf{T}_E(\mathbf{P}_\Lambda^k(x)) - \mathbf{T}_E(\mathbf{P}_\Lambda^k(w))\right) &= \lim_{l\to\infty}\left(\int_0^{T^l(x)} dt - \int_0^{T^l(w)} dt\right)\\ &= \lim_{l\to\infty}\left(\int_{c^l(x)} \theta - \int_{c^l(w)} \theta\right)\end{aligned} \tag{8.8}$$

with the path $c^l(z) : [0, T^l(z)] \to \boldsymbol{\Sigma}_E$, $t \mapsto \boldsymbol{\Phi}_E^t(z)$, since

$$\int \theta = \int p_i dq^i = \int p_i(t)\dot{q}^i(t)dt = \int dt$$

on the unit tangent bundle $\boldsymbol{\Sigma}_E$ of $(\mathbf{M}, \mathbf{g}_E)$.

Let $c(x,w)$ be the oriented segment of the curve $V(f_x^+) \subset C_l$ going from x to w, and let $D_1^l \subset \boldsymbol{\Sigma}_E$ be the two-cell parametrized by

$$D_1^l = \{\boldsymbol{\Phi}_E^t(z) \mid z \in c(x,w), 0 \leq t \leq T^l(z)\}.$$

Then $\partial D_1^l = c^l(x) - c^l(w) - c(x,w) + \mathbf{P}^l(c(x,w))$. The length of the curve $\mathbf{P}^l(c(x,w))$ goes to zero as $l \to \infty$, see Fig. 8.1 b).

Then by Stokes' formula we are allowed to rewrite (8.8) in the form

$$\begin{aligned}
&\sum_{k=0}^{\infty}\left(\mathbf{T}_E(\mathbf{P}_\Lambda^k(x)) - \mathbf{T}_E(\mathbf{P}_\Lambda^k(w))\right)\\
&\quad = \lim_{l\to\infty}\left(\int_{c^l(x)}\theta - \int_{c^l(w)}\theta - \int_{\partial D_1^l}\theta + \int_{D_1^l} d\theta\right)\\
&\quad = \int_{c(x,w)}\theta - \lim_{l\to\infty}\int_{D_1^l}\omega_E = \int_{c(x,w)}\theta,
\end{aligned} \tag{8.9}$$

since $\lim_{l\to\infty}\int_{\mathbf{P}^l(c(x,w))}\theta = 0$ and since $\int_{D_1^l}\omega_E = 0$, the last equality being due to the fact that D_1^l lies in the stable manifold of the orbit of x which is Lagrangian by Thm. 5.3.30 of [1].

In a similar manner we introduce for the remaining pairs of points in (8.6) and (8.7) oriented segments $c(w,y) \subset W(f_y^-)$, $c(y,v) \subset V(f_y^+)$, and $c(v,x) \subset W(f_x^-)$, and we end up with the equation

$$\begin{aligned}
\Delta T(x,y) &= \int_{c(x,w)}\theta + \int_{c(w,y)}\theta + \int_{c(y,v)}\theta + \int_{c(v,x)}\theta\\
&= \int_{\partial F}\theta = \int_F d\theta = -\int_F \omega_E,
\end{aligned}$$

proving (8.5). □

Proposition 8.3 *For $n \geq 3$ and E large, the geodesic flow Φ_E^t, restricted to the set $\mathbf{b}_E \subset \Sigma_E$ of bounded orbits, is weakly mixing w.r.t. the measure of maximal entropy. Moreover, the number $\mathbf{N}_E(T)$ of closed geodesics of minimal period smaller than T is asymptotic to*

$$\mathbf{N}_E(T) \sim \frac{\exp(h_{\mathrm{top}}T)}{h_{\mathrm{top}}T}, \tag{8.10}$$

with $h_{\mathrm{top}} \equiv h_{\mathrm{top}}(\Phi_E^1)$ analysed in Proposition 7.5.

Proof. In [6], Bowen analysed the non-wandering set Ω of Axiom A flows Φ^t, see Definition 7.3. By the spectral decomposition theorem that set decomposes into the disjoint union

$$\Omega = \Omega_1 \cup \ldots \cup \Omega_k$$

of closed invariant sets Ω_i on which Φ^t is *topologically transitive*, that is, there exists some $x_i \in \Omega_i$ whose orbit $\{\Phi^t(x) \mid t \in \mathbb{R}\}$ is dense in Ω_i. These Ω_i are called *basic sets.*

For these basic sets, Bowen proved the following alternative:

"*Either* $\Phi^t|_{\Omega_i}$ has a non-constant continuous eigenfunction $g : \Omega_i \to \mathbb{C}$, *i.e.* for some $a \neq 0$

$$g(\Phi^t(x)) = e^{iat}g(x), \quad \forall x \in \Omega_i, \forall t \in \mathbb{R}, \tag{8.11}$$

or $\Phi^t|_{\Omega_i}$ is weakly mixing w.r.t. the measure of maximal entropy."

Now in our case, we do not formally have an Axiom A flow, since the energy shell $\boldsymbol{\Sigma}_E$ is not compact. Nevertheless, the non-wandering set

$$\Omega(\boldsymbol{\Phi}_E) = \mathbf{b}_E$$

for E large, by Lemma 7.4.

Thus $\Omega(\boldsymbol{\Phi}_E)$ is compact and carries a hyperbolic structure, again by Prop. 6.9. For $n = 2$, by Prop. 6.9, Ω splits into two basic sets each consisting of a closed geodesic flow line, whereas for $n \geq 3$, Ω consists of one basic set.

All constructions in the papers of Bowen and of Parry and Pollicott are local w.r.t. the basic sets. Thus they extend to our situation.

In [37] it is shown that for basic sets of weakly mixing Axiom A flows formula (8.10) holds true. So we can prove the proposition by showing the non-existence of a continuous non-constant eigenfunction $g : \mathbf{b}_E \to \mathbb{C}$ in the sense of (8.11).

Assume to the contrary that such a g exists, with frequency $a \neq 0$. Then $|g| \equiv \text{const} \neq 0$, by continuity of g, eq. (8.11) and existence of a dense orbit. We shall derive a contradiction, starting with a point x in a Poincaré section, that is, $x \in \mathbf{b}_E \cap C_l$ for some $l \in \{1, \ldots, 2(n-1)\}$.

By continuity, $g(z) = g(x)$ for all $z \in W^s(x)$ or $z \in W^u(x)$, see (8.2).

Let y be contained in the same Poincaré section C_l. Then the definition (8.3) of the point $\langle x, y \rangle$ implies that

$$g(y) = g(\langle x, y \rangle) = g(\boldsymbol{\Phi}_E^{u(x,y)}(x)) \text{ and } g(x) = g(\langle y, x \rangle) = g(\boldsymbol{\Phi}_E^{u(y,x)}(y)),$$

see Fig. 8.1 a).

On the other hand,

$$g(\boldsymbol{\Phi}_E^{u(x,y)}(x)) = \exp(iau(x,y))g(x) \text{ and } g(\boldsymbol{\Phi}_E^{u(y,x)}(y)) = \exp(iau(y,x))g(y)$$

which implies

$$g(x) = \exp(ia(u(x,y) + u(y,x)))g(x)$$

so that

$$\Delta T(x,y) = u(x,y) + u(y,x) = 2\pi k/a \quad \text{for some } k \in \mathbb{Z},$$

since $a \neq 0$ and $g(x) \neq 0$ by assumption.

Now we construct a sequence of $y_k \in C_l$, $k \in \mathbb{N}$, with $\Delta T(x, y_k) \neq 0$ and

$$\lim_{k \to \infty} \Delta T(x, y_k) = 0. \tag{8.12}$$

This then implies $a = 0$, contradicting our assumption.

Let $f := \mathcal{H}^{-1}(x)$ be the symbol sequence of x. We set $y_k := \mathcal{H}(f_k)$, with $f_k \in \mathbf{X}$ given by

$$f_k(l) := f(l) \text{ for } |l| \neq k \quad \text{and } f_k(k) \neq f(k), f_k(-k) \neq f(-k).$$

This is always possible since the entries of the matrix $\mathcal{M}^2$ are larger or equal to two for $n \geq 3$:

$$\left(\mathcal{M}^2\right)_{ik} = 2(n-2) + \delta_{ik}.$$

We estimate the desynchronization time $\Delta T(x, y_k)$ using Lemma 8.2. Due to our construction, the areas of the regions $F_k \subset C_l$ are non-zero which implies $\Delta T(x, y_k) = \int_{F_k} \omega_E \neq 0$. On the other hand, y_k converges to x implying (8.12). □

Remark 8.4 Equivalently,

$$\pi_E(x) := \mathbf{N}_E(h_{\text{top}}^{-1} \ln x) \sim \frac{x}{\ln x}. \tag{8.13}$$

Thus, Proposition 8.3 has the form of a prime number theorem. Therefore, it is natural to examine the *zeta function*

$$\zeta(s) := \prod_{\tau \text{ closed orbit}} (1 - \exp(-s\lambda(\tau)))^{-1}$$

where $\lambda(\tau)$ denotes the minimal period of the periodic flow line τ. In our case, by Thm. 1 of [37], $\zeta(s)$ has a nowhere vanishing analytic extension to a neighbourhood of the closed half plane $\{s \in \mathbb{C} \mid \Re(s) \geq h_{\text{top}}(\Phi_E^1)\}$, except for a simple pole at $s = h_{\text{top}}(\Phi_E^1)$. Moreover, the zeta function has an unusually simple form in our case since $\mathbf{b}_E$ is homeomorphic to the space $\mathbf{X}_E$ (technically, there are no 'auxiliary shift suspensions' in the sense of [37]).

We will not study these zeta functions in more detail.

Unfortunately, we only get a slightly weaker result on the closed orbits of the flow Φ_E^t on Σ_E, compared to Prop. 8.3:

Theorem 8.5 *For $n \geq 3$ and E large, the Hamiltonian flow Φ_E^t, restricted to the set $b_E \subset \Sigma_E$ of bounded orbits, is ergodic w.r.t. the measure of maximal entropy. Moreover there is $C > 1$ such that the number $N_E(T)$ of closed orbits of minimal period smaller than T verifies the estimate*

$$C^{-1} \frac{\exp(h_{\text{top}} T)}{h_{\text{top}} T} \leq N_E(T) \leq C \frac{\exp(h_{\text{top}} T)}{h_{\text{top}} T} \tag{8.14}$$

for T large and $h_{\text{top}} \equiv h_{\text{top}}(\Phi_E^1)$ analysed in Theorem 7.6.

Proof. As above, this follows from Theorem 2 of [37]. □

Remarks 8.6

1. For ergodic geodesic flows on *compact* manifolds (not homeomorphic to the two-torus) a beautiful theorem of Arnol'd says that neither the geodesic flow nor any flow obtained from it by a continuous change of time can have such a nontrivial eigenfunction. The proof of that theorem is of topological nature, see, e.g. Anosov [2], §23. Unfortunately, in our situation, no direct analog of the theorem of Arnol'd exists, for the following reason:

By Prop. 6.9 the geodesic flow $\boldsymbol{\Phi}_E^t$, restricted to $\mathbf{b}_E$, is conjugate to the suspension flow $\boldsymbol{\sigma}_E^t$. Therefore, by a continuous change in the time parametrization, we can make the flow conjugate to a constant time suspension, which is certainly *not* mixing.

2. Furthermore, as the following counterexample shows, there is no general geometric principle comparable to Lemma 8.2 for Hamiltonian functions of the form 'kinetic + potential' which could provide nonzero desynchronization times ΔT.

   Consider the motion on the two-torus $\mathbb{T}^2$ given by the Hamiltonian $H : T^*\mathbb{T}^2 \to \mathbb{R}$,

   $$H(\vec{q},\vec{p}) := H_1(q_1,p_1) + \tfrac{1}{2}p_2^2, \quad \text{with } H_1(q_1,p_1) := \tfrac{1}{2}p_1^2 + \cos(4\pi q_1)$$

   for an energy $H \equiv E > 1$. Then the orbits

   $$(\vec{q}_I(t), \vec{p}_I(t)) := (0, \sqrt{2E}t, 0, \sqrt{2E})$$

   and

   $$(\vec{q}_{II}(t), \vec{p}_{II}(t)) := (2\pi, \sqrt{2E}t, 0, \sqrt{2E})$$

   are hyperbolic. Their stable and unstable manifolds are contained in the set

   $$\{(\vec{q},\vec{p}) \in T^*\mathbb{T}^2 \mid H(\vec{q},\vec{p}) = E, H_1(q_1,p_1) = 1\}.$$

   The desynchronization time $\Delta T(x_I, x_{II}) = 0$ for $x_i := (\vec{q}_i(0), \vec{p}_i(0))$, since the motions in the 1- and 2-direction separate (but observe that the associated geodesic motion for the Jacobi metric does *not* separate).

   So a proof of a non-zero desynchronization time for the flow Φ^t generated by our Hamiltonian function (1.1) must be of analytic, not of geometric or topological nature.

9. Fractional Dimension

In this chapter we shall estimate the fractional dimension of the set b_E of bounded orbits of energy E, for E large. This quantity, being of interest in its own right, governs the measure of those scattering orbits which have a large time delay. In our semiclassical analysis of the quantum mechanical problem, the fractional dimension of the bounded orbits will determine a bound for the number of resonance poles near the real energy axis.

Besides the well-known dimension introduced by Hausdorff and Besicovitch, there exist other definitions of the fractional dimension.

Definition 9.1 Let (X, d) be a metric space and $U \subset X$, $U \neq \emptyset$. The *diameter* $\mathrm{diam}(U)$ *of* U is given by

$$\mathrm{diam}(U) := \sup\{d(x,y) \mid x, y \in U\}.$$

For $E \subset X$ and $\delta > 0$ let

$$\mathcal{H}^s_\delta(E) := \inf\left\{\sum_{i=1}^{\infty}(\mathrm{diam}(U_i))^s \mid E \subset \cup_{i=1}^{\infty} U_i, \mathrm{diam}(U_i) \leq \delta\right\}.$$

The *Hausdorff s-dimensional outer measure of* E, $\mathcal{H}^s(E)$, is then defined by

$$\mathcal{H}^s(E) := \sup_{\delta > 0} \mathcal{H}^s_\delta(E) = \lim_{\delta \searrow 0} \mathcal{H}^s_\delta(E). \tag{9.1}$$

The *Hausdorff dimension* of E is given by

$$\dim_H(E) := \sup\{s \in \mathbb{R} \mid \mathcal{H}^s(E) = \infty\}. \tag{9.2}$$

The important property of $\mathcal{H}^s$ is that it defines a measure on, say, the Borel sets. Therefore, Hausdorff dimension is not only *monotone*, that is

$$E_1 \subset E_2 \Rightarrow \dim_H(E_1) \leq \dim_H(E_2), \tag{9.3}$$

but also *σ-stable*:

$$\dim_H(\cup_i E_i) = \sup_i \dim_H(E_i). \tag{9.4}$$

The quantity $\dim_K$ defined below runs under different names in the literature: the *metric dimension* of A. Kolmogorov, the *logarithmic density* of C. Tricot, or the *entropy dimension* of J. Hawkes, see C. Tricot, Jr [45].

$\dim_K$ will be used in [24] to derive a semiclassical upper bound for the number of quantum mechanical resonances.

Definition 9.2 Let (X, d) be a metric space and $E \subset X$, $U \neq \emptyset$. We denote by $Q(r)$ the set of open balls of radius $\leq r$. Let

$$\begin{aligned} \tilde{\mathcal{H}}_\delta^s(E) &:= \sup\{\sum_{i=1}^{\infty}(\mathrm{diam}(U_i))^s \mid U_i \in Q(\delta), \\ &\quad \inf\{d(u,e) \mid u \in U_i, e \in E\} = 0, U_i \cap U_k = \emptyset \text{ for } i \neq k\} \end{aligned}$$

and $\tilde{\mathcal{H}}^s(E) := \sup_{\delta>0} \tilde{\mathcal{H}}_\delta^s(E)$.

The dimension $\dim_K(E)$ is given by

$$\dim_K(E) := \sup\{s \in \mathbb{R} \mid \tilde{\mathcal{H}}^s(E) = \infty\}.$$

$\dim_K$ is monotone (see (9.3)), and

$$\dim_K(E_1 \cup E_2) = \max\{\dim_K(E_1), \dim_K(E_2)\},$$

but it is not σ-stable, since $\dim_K$ is invariant under closure. Furthermore, for all E,

$$\dim_H(E) \leq \dim_K(E). \tag{9.5}$$

For example, the triadic Cantor set $E \subset \mathbb{R}$ has

$$\dim_H(E) = \dim_K(E) = \frac{\ln 2}{\ln 3} = 0.6309\cdots.$$

On the other hand, the set $\mathbb{Q} \subset \mathbb{R}$ of rational numbers has dimensions

$$\dim_H(\mathbb{Q}) = 0, \qquad \dim_K(\mathbb{Q}) = 1.$$

Typically it is much easier to estimate fractional dimensions from above than from below, the situation being comparable to eigenvalue estimates in quantum mechanics. The reason is simply that *any* prescription for a δ-covering of the set E leads to an upper bound.

The proof of our lower estimate of $\dim_H(\mathbf{b}_E)$ is not based on a convexity argument as is the usual proof for the triadic Cantor set, since $\mathbf{b}_E$ is more irregular. Instead, we use a technique which generalizes a method applied in the proof of Thm. 8.6 in the beautiful book [10] of Falconer.

Proposition 9.3 *For $n \geq 2$ and E large, the set $\mathbf{b}_E$ of bounded E-geodesic flow lines has dimensions bounded by*

$$1 + \frac{2\ln(2n-3)}{J_u \ln E} \leq \dim_H(\mathbf{b}_E) \leq \dim_K(\mathbf{b}_E) \leq 1 + \frac{2\ln(2n-3)}{J_l \ln E}. \tag{9.6}$$

Proof. We analyse the fractional dimension of $\mathbf{b}_E$ by considering its intersection $\Lambda_l := \mathbf{b}_E \cap C_l$ with the Poincaré sections C_l, $l \in \{1, \ldots, 2(n-1)\}$. Set $k_0 := l$. Then for all $m \in \mathbb{N}$ Λ_l is contained in the union of the 'rectangles' $W(k_{-m}, \ldots, k_0) \cap V(k_0, \ldots, k_m) \subset C_{k_0}$, with $(k_{-m}, \ldots, k_m)$ admissible.

On C_l we have the distance function dist derived from the Riemannian metric on the energy shell Σ_E. W.r.t. that metric, for $\delta > 0$ let N_δ be the smallest number of closed balls in C_l of diameter δ covering Λ_l. By a compactness argument, in a compact region in C_l containing an open neighbourhood of Λ_l, the metric is uniformly equivalent to the Euclidean metric derived from the chart $(l, \cos\alpha)$ on C_l defined in (6.6). Therefore, by Corollary 2 of [45], we have

$$\dim_K(\Lambda_l) = \limsup_{\delta \searrow 0} \frac{\ln(N_\delta)}{-\ln \delta}. \tag{9.7}$$

We use (9.7) to derive the rightmost inequality in (9.6). Thus we estimate N_δ from above for small values of $\delta > 0$.

If $m \geq m_0$, we know from the estimates (6.19) and (6.21) of Lemma 6.7 that the maximal diameter d_m of the 'rectangles' $W(k_{-m}, \ldots, k_0) \cap V(k_0, \ldots, k_m) \subset C_{k_0}$ is bounded by

$$d_m \leq c' \exp(-J_l m \ln E), \qquad \text{with } c' := 2\left(1 + \frac{E^2}{s_l s_{\min}^2}\right). \tag{9.8}$$

There are $(2n-3)^{2m}$ such 'rectangles' for $k_0 = l$ fixed.

For $\delta > 0$ small and $m := [|\ln(\delta/c')| / (J_l \ln E)] + 1$ we can cover each 'rectangle' by a ball of radius δ. Thus

$$\ln(N_\delta) \leq 2m \cdot \ln(2n-3) \leq 2\left(\frac{|\ln(\delta/c')|}{J_l \ln E} + 1\right) \ln(2n-3),$$

so that

$$\dim_K(\Lambda_l) \leq \frac{2\ln(2n-3)}{J_l \ln E}. \tag{9.9}$$

Now we derive a lower bound for the Hausdorff dimension of $\Lambda_l \subset C_l$. We set

$$s := \frac{2\ln(2n-3)}{J_u \ln E}. \tag{9.10}$$

and prove the existence of a $C = C(E) > 0$ for which all coverings $\{U_i\}_{i \in \mathbb{N}}$ of Λ_l meet the inequality

$$\sum_{i \in \mathbb{N}} (\mathrm{diam}(U_i))^s \geq C. \tag{9.11}$$

(9.11) implies $\dim_H(\Lambda_l) \geq s$, since

$$\dim_H(\Lambda_l) \equiv \sup\{t \in \mathbb{R} \mid \mathcal{H}^t(\Lambda_l) = \infty\} = \inf\{t \in \mathbb{R} \mid \mathcal{H}^t(\Lambda_l) = 0\}.$$

By (9.1), we may confine our considerations to coverings with small diameter $\delta > 0$. Let $B(\delta) \subset C_l$ be a ball of radius δ. Our first task is to estimate from above the number N of 'rectangles' $W(k_{-m},\ldots,k_0)\cap V(k_0,\ldots,k_m)$ intersected by $B(\delta)$, with

$$m := \left[\frac{|\ln\delta|}{J_u \ln E}\right]. \tag{9.12}$$

Obviously, $N \le C_w C_v$, where C_w (C_v) is the maximal number of strips $W(k_{-m},\ldots,k_0)$ ($V(k_0,\ldots,k_m)$) intersected by the ball $B(\delta)$. By eqs. (6.22) and (6.21) of Lemma 6.7, the minimal distance between a point x_u on the upper boundary of a strip in $W(k_{-m},\ldots,k_0)$ and a point x_l on the lower boundary is

$$d(x_u,x_l) \ge \exp(-J_u m \ln E)\cdot c'' \le \delta c''$$

with $c'' := 1/\sqrt{1+(s_u E)^2}$, the last inequality following from (9.12).

Thus $C_w \le 2+1/c''$, since $(C_w-2)d(x_u,x_l)\le\delta$. Clearly, $C_v \le 2+1/c''$, too, so that

$$N \le C_w C_v \le (2+1/c'')^2,$$

the estimate being independent of $\delta > 0$.

We want to estimate the 'proportion' of Λ_l covered by a ball $B(\delta)$ of radius δ. To formalize that notion, we introduce a probability measure μ_l on C_l concentrated on Λ_l (that is, $\mu_l(\Lambda_l)=1$).

First we introduce the cylinder measure μ on the space $\mathbf{X}$ of admissible sequences by setting

$$\mu(\{f\in\mathbf{X} \mid (f_m,\ldots,f_{m+k})=(a_0,\ldots,a_k)\}) := \left(2(n-1)(2n-3)^k\right)^{-1}$$

for $m\in\mathbb{Z}$, $k\in\mathbb{N}$ and $(a_0,\ldots,a_k)$ admissible, and extending μ to the σ-algebra generated by these cylinder sets.

Then we define μ_l for a Borel set $U\subset C_l$ by

$$\mu_l(U) := 2(n-1)\cdot\mu(\mathcal{H}^{-1}(U\cap\Lambda_l)).$$

The proportion of Λ_l covered by a ball $B(\delta)$ is then bounded by

$$\begin{aligned}\mu_l(B(\delta)\cap\Lambda_l) &\le 2(n-1)\frac{N}{2(n-1)}(2n-3)^{-2m}\\ &\le N(2n-3)^2\cdot\delta^s,\end{aligned}$$

with $m\equiv m(\delta)$ given by (9.12) and s defined in (9.10).

Let $\{U_i\}_{i\in\mathbb{N}}$ be an arbitrary cover of Λ_l. We may then cover Λ_l by balls $\{B_i\}_{i\in\mathbb{N}}$ with $\mathrm{diam}(B_i)\le 2\mathrm{diam}(U_i)$, so that

$$\begin{aligned}\sum_{i\in\mathbb{N}}(\mathrm{diam}(U_i))^s &\ge 2^{-s}\sum_{i\in\mathbb{N}}(\mathrm{diam}(B_i))^s\\ &\ge C\sum_{i\in\mathbb{N}}\mu_l(B_i\cap\Lambda_l)\ge C\mu_l(\Lambda_l)=C,\end{aligned}$$

with $C := ((2n-3)(2+1/c''))^{-2}$, proving (9.11).

We obtain the estimates for the dimensions of $\mathbf{b}_E$ by noticing that, near C_l, $\mathbf{b}_E$ is diffeomorphic to $\Lambda_l \times (0,1)$, since the flow $\boldsymbol{\Phi}_E^t$ is transversal to C_l near Λ_l, and by linearization of the flow.

The interval $(0,1)$ is *regular* in the sense that

$$\dim_H((0,1)) = \dim_K((0,1)) = 1.$$

In [45], Tricot shows that for F regular,

$$\dim_H(E \times F) = \dim_H(E) + \dim_H(F),$$

and the proof of the equality $\dim_K(E \times F) = \dim_K(E) + \dim_K(F)$ is implicitly contained in the proof of his Theorem 3.

We can find a finite cover of $\mathbf{b}_E$ by sets of the above form. Then stability of $\dim_H$ and $\dim_K$ implies eq. (9.6) of the proposition. □

The dimensions of $b_E \subset \Sigma_E$ obey the same bounds as on the covering space:

Theorem 9.4 *For $n \geq 2$ and E large, the set b_E of bounded orbits of Φ_E^t has dimensions bounded by*

$$1 + \frac{2\ln(2n-3)}{J_u \ln E} \leq \dim_H(b_E) \leq \dim_K(b_E) \leq 1 + \frac{2\ln(2n-3)}{J_l \ln E}. \tag{9.13}$$

Proof. We would like to derive (9.13) from (9.6), using the local homeomorphism $\pi_E : \boldsymbol{\Sigma}_E \to \Sigma_E$ and the relation $\pi_E(\mathbf{b}_E) = b_E$. Unfortunately $\dim_H$ and $\dim_K$ are not invariants under homeomorphisms but only under diffeomorphisms, and π_E is not a local diffeomorphism due to the presence of the points projecting to the positions $\vec{s}_l$ of the nuclei.

Nevertheless these circles are of codimension two in Σ_E, and the flow Φ_E^t is transversal w.r.t. them. Therefore we can make the following construction. We use a finite cover $B_1, \ldots, B_m \subset \Sigma_E$ of $b_E \subset \cup_{l=1}^m B_l$ by compact balls B_l of small diameters.

For those B_l with $B_l \cap \pi^{-1}(\vec{s}_k) = \emptyset$, we know that $\dim_H(B_l \cap b_E) = \dim_H(\pi_E^{-1}(B_l \cap b_E))$ (and similarly for $\dim_K$).

For the case $B_l \cap \pi^{-1}(\vec{s}_k) \neq \emptyset$ we estimate $\dim_H(B_l \cap b_E)$ by exploiting the flow invariance of the dimension:

$$\dim_H(B_l \cap b_E) = \dim_H(\Phi_E^t(B_l \cap b_E))$$

with $t \in \mathbb{R}$ chosen so that $\Phi_E^t(B_l \cap b_E) \cap \pi^{-1}(\vec{s}_k) = \emptyset$. □

10. Time Delay

In this chapter we relate the time delay τ of the scattering orbits to the structure of the motion near the bounded orbits.

In [34], Narnhofer gave a definition of that quantity for the case of a motion in a short-range potential V (with $|\nabla V(\vec{q})| = \mathcal{O}(|\vec{q}|^{-2-\epsilon})$). The time delay is the difference between the time spent by a particle in a ball of radius R, and the time spent by the corresponding incoming free particle, in the limit $R \to \infty$.

In that case, for a point $(\vec{q}_0, \vec{p}_0)$ on a scattering orbit the limits

$$\vec{p}^{\pm} := \lim_{t\to\pm\infty} \vec{p}(t) \quad \text{and} \quad \vec{q}^{\pm} := \lim_{t\to\pm\infty} (\vec{q}(t) - \vec{p}(t)t)$$

are well-defined, and the time delay is given by

$$\tau(\vec{q}_0, \vec{p}_0) = \frac{\vec{q}^{-} \cdot \vec{p}^{-}}{|\vec{p}|^2} - \frac{\vec{q}^{+} \cdot \vec{p}^{+}}{|\vec{p}|^2}.$$

We remark that due to the simple form of the free time evolution comparison with the corresponding *outgoing* free particle gives the same value of the time delay.

This differs from our case, since we compare the motion Φ^t with the motion Φ^t_∞ generated by a purely Coulombic potential, cf. (2.19). In order to keep the time reversal symmetry, we therefore average over the incoming and outgoing asymptote.

The time delay τ is then a function of the scattering initial conditions $s \in P$ introduced in Def. 2.4.

Definition 10.1 The *time delay* $\tau : s \to \mathbb{R}$ of a scattering state $x \in s$ is given by

$$\tau(x) := \lim_{R\to\infty} \int_{\mathbb{R}} \Big(\sigma(R) \circ \Phi^t(x) - \frac{1}{2}(\sigma_\infty(R) \circ \Phi^t_\infty \circ \Omega^+_*(x) + \sigma_\infty(R) \circ \Phi^t_\infty \circ \Omega^-_*(x))\Big)\, dt, \tag{10.1}$$

where $\sigma(R) : P \to \{0,1\}$ and $\sigma_\infty(R) : P_\infty \to \{0,1\}$ are the characteristic functions $\sigma(R)(x) := \theta(R - |\eta(x)|)$ and similarly for $\sigma_\infty(R)$.

So the time delay is the difference between the time spent by the trajectory $\Phi^t(x)$ inside a ball $|\vec{q}| \le R$ in the configuration plane, and the mean of the

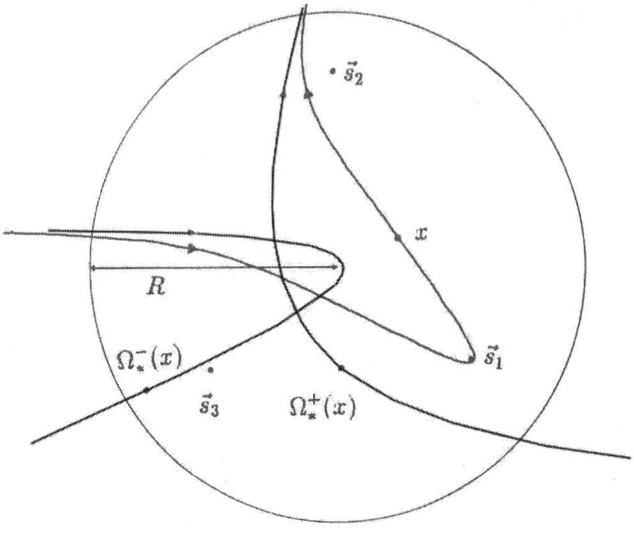

Fig. 10.1. On the definition of time delay.

times spent by the Kepler solutions $\Phi_\infty^t(\Omega_*^\pm(x))$ in the same ball, in the limit $R \to \infty$, see Fig. 10.1.

Formally, we see that τ is translation invariant, *i.e.* $\tau \circ \Phi^t = \tau$, using (2.27). Moreover, our definition makes τ invariant w.r.t. time reversal (which is an antisymplectic transformation of phase space mapping $(\vec{q}, \vec{p})$ onto $(\vec{q}, -\vec{p})$).

Of course, we must show that the above limit really exists.

Lemma 10.2 *The time delay* τ *is well-defined, and for a scattering state* $x \in s$

$$\begin{aligned}
&\tau(x) &(10.2)\\
&= \tfrac{1}{2}\lim_{R\to\infty} \textstyle\int_{\mathbb{R}} \sigma^+ \circ \Phi^t(x) \cdot (\sigma_\infty(R) \circ \Omega_*^+ - \sigma_\infty(R) \circ \Omega_*^-)(\Phi^t(x))dt \\
&+ \tfrac{1}{2}\lim_{R\to\infty} \textstyle\int_{\mathbb{R}} \sigma^- \circ \Phi^t(x) \cdot (\sigma_\infty(R) \circ \Omega_*^- - \sigma_\infty(R) \circ \Omega_*^+)(\Phi^t(x))dt,
\end{aligned}$$

with $\sigma^\pm((\vec{q}, \vec{p})) := \theta(\pm\vec{q} \cdot \vec{p})$.

Moreover, there exists a decreasing function $C(E) > 0$ *of the energy* E *with* $\lim_{E\to\infty} C(E) = 0$ *such that for all* $x \in s$ *whose orbits never enter the ball of radius* R_{vir}, $|\tau(x)| < C(H(x))$.

Proof. The inverse Møller transformations conjugate the flows: $\Omega_*^\pm \circ \Phi^t = \Phi_\infty^t \circ \Omega_*^\pm$. We write the integrand of (10.1) in the form

$$\begin{aligned}
&\sigma(R) \circ \Phi^t(x) - \tfrac{1}{2}\left(\sigma_\infty(R) \circ \Phi_\infty^t \circ \Omega_*^+(x) + \sigma_\infty(R) \circ \Phi_\infty^t \circ \Omega_*^-(x)\right) \\
&= A^+ + A^-
\end{aligned}$$

with

$$A^\pm := \tfrac{1}{2}\left(\sigma(R) \circ \Phi^t(x) - \sigma_\infty(R) \circ \Phi_\infty^t \circ \Omega_*^\pm(x)\right).$$

Then we evaluate $A^{\pm}$ by noticing that $A^{\pm} = (\sigma^{+} \circ \Phi^{t}(x) + \sigma^{-} \circ \Phi^{t}(x))A^{\pm}$ almost everywhere.

For x fixed and R large, $\sigma^{+} \circ \Phi^{t}(x) \cdot A^{\pm}$ is non-zero only for large positive values of t. Then

$$\lim_{t\to\infty} \operatorname{dist}\left(\Phi^{t}(x), \Phi^{t}_{\infty} \circ \Omega^{+}_{*}(x)\right) = \lim_{t\to\infty} \operatorname{dist}\left(\Phi^{t}(x), \Omega^{+}_{*} \circ \Phi^{t}(x)\right) = 0$$

by the definition of the inverse Møller transformation $\Omega^{\pm}_{*}$. Thus the term $\frac{1}{2}\sigma^{+} \circ \Phi^{t}(x) \cdot A^{+}$ in the integrand of (10.1) does not contribute.

On the other hand, $\sigma^{-} \circ \Phi^{t}(x) \cdot A^{\pm}$ is non-zero only for large negative values of t. In that case,

$$\lim_{t\to-\infty} \operatorname{dist}\left(\Phi^{t}(x), \Phi^{t}_{\infty} \circ \Omega^{-}_{*}(x)\right) = \lim_{t\to-\infty} \operatorname{dist}\left(\Phi^{t}(x), \Omega^{-}_{*} \circ \Phi^{t}(x)\right) = 0$$

so that

$$\begin{aligned} &\lim_{R\to\infty} \tfrac{1}{2} \int_{\mathbb{R}} (\sigma(R) \circ \Phi^{t}(x) - \sigma_{\infty}(R) \circ \Phi^{t}_{\infty} \circ \Omega^{+}_{*}(x))\, dt \\ = &\lim_{R\to\infty} \tfrac{1}{2} \int_{\mathbb{R}} \sigma^{-} \circ \Phi^{t}(x) \left(\sigma_{\infty}(R) \circ \Omega^{-}_{*} - \sigma_{\infty}(R) \circ \Omega^{+}_{*}\right) \Phi^{t}(x) dt. \end{aligned}$$

Clearly that limit is well-defined. In fact, it involves only estimates on Kepler hyperbolae since

$$\begin{aligned} &\left(\sigma_{\infty}(R) \circ \Omega^{-}_{*} - \sigma_{\infty}(R) \circ \Omega^{+}_{*}\right) \Phi^{t}(x) \\ = \; &\sigma_{\infty}(R) \circ \Phi^{t}_{\infty} \circ \Omega^{-}_{*}(x) - \sigma_{\infty}(R) \circ \Phi^{t}_{\infty} \circ \Omega^{+}_{*}(x). \end{aligned}$$

A similar reasoning, applied to A^{-}, leads to (10.2).

The time delay is invariant w.r.t. Φ^{t}. Therefore instead of estimating $\tau(x)$ from above for all those scattering states $x \in s$ of a given energy $H(x) = E$ which never enter the interaction zone of radius R_{vir}, we may equally well estimate $\tau(x)$ only for those $x = (\vec{q}, \vec{p})$ with $\vec{q} \cdot \vec{p} = 0$. Then by the virial inequality $|\vec{q}|$ is the minimal distance from the origin for the whole orbit. It is clear that $\tau(x) \to 0$ for $|\vec{q}| \to \infty$, since then $\operatorname{dist}(x, \operatorname{Id} \circ \Omega^{\pm}_{*}(x)) \to 0$.

Moreover, a rescaling of momentum and time shows that as the energy $E = H(\vec{q}, \vec{p})$ goes to infinity, for $\vec{q} \cdot \vec{p} = 0$ and $|\vec{q}| > R_{\text{vir}}$, $\tau(\vec{q}, \vec{p}) \to 0$. □

Remark 10.3 The time delay is a continuous function on the set $s \in P$ of scattering states. This is a consequence of the continuity of the Møller transformations (Prop. 2.7) and (10.2), combined with the explicit form (2.16) of the relation between time and radius for the Keplerian motion.

Moreover, if the potential V meets the set of equations (2.35), then by Prop. 2.9 the $\Omega^{\pm}$ are smooth canonical transformations. Using this fact we conclude that in that situation τ is smooth.

We would like to express time delay as a function of the asymptotic momenta $\vec{p}^{\pm}$ and angular momenta $L^{\pm}$ (2.28) of the scattering orbits. This is possible since $\tau \circ \Phi^{t} = \tau$ for all times t.

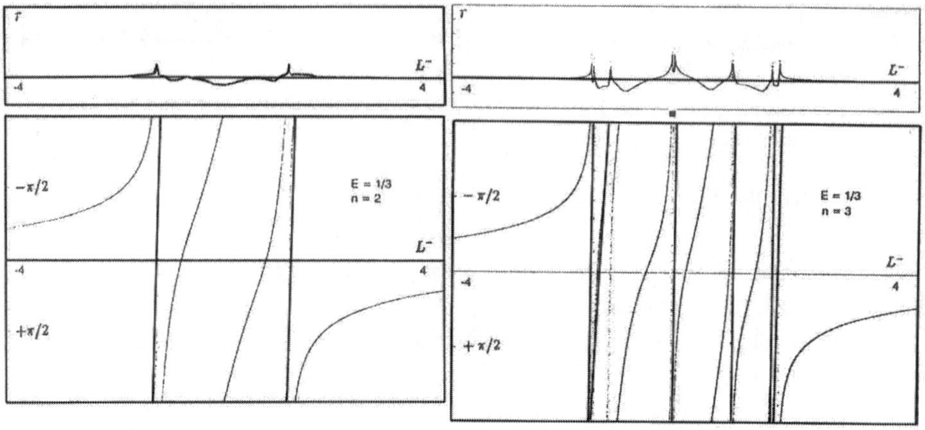

a b

Fig. 10.2. Time delay (above) and scattering angle φ^+ (below) as functions of the initial angular momentum L^- for a) $n = 2$, b) $n = 3$ centers

We are a bit more general in defining the time delay

$$\tau^\pm : \mathbb{R}^+ \times S^1 \times \mathbb{R} \to \mathbb{R} \cup \{\infty\}$$

as a function of energy E, asymptotic scattering angle $\varphi^\pm$ and angular momentum ($p_1^\pm = \sqrt{2E}\cos\varphi^\pm$, $p_2^\pm = \sqrt{2E}\sin\varphi^\pm$):

$$\tau^\pm\left(H(x), \varphi^\pm(x), L^\pm(x)\right) := \tau(x) \qquad \text{for } x \in s.$$

From asymptotic completeness (Corollary 2.8, 2.), we see that this relation fixes $\tau^\pm$ on its domain of definition up to a set of measure zero. Nevertheless, the symmetric difference $(s^+ \cup s^-) \setminus s \neq \emptyset$ in general (certainly for $n \geq 2$, and sometimes for $n = 1$ in the low-energy range). In other words, there exist orbits which are captured by the interaction zone for $t \to \mp\infty$. For points $x \in s^\pm \setminus s$ on those orbits we set $\tau^\pm(H(x), \varphi^\pm(x), L^\pm(x)) := \infty$.

In Fig. 10.2 a) and b) we show a numerical plot of the time delay for fixed energy E and initial angle φ^- as a function of the initial angular momentum L^- for $n = 2$ and $n = 3$, using the potentials $V(\vec{q}) := -\sum_{l=1}^n Z/|\vec{q} - \vec{s}_l|$ with $Z := 1/n$. The positions of the nuclei were centered at $\sum_{l=1}^n \vec{s}_l = \vec{0}$, and their mutual distance was $|\vec{s}_l - \vec{s}_k| = 1 - \delta_{lk}$. We do not reproduce a similar picture for $n = 1$ since there $\tau \equiv 0$ for a purely Coulombic potential at $\vec{s}_1 = \vec{0}$.

For $n = 2$, the time delay is seen to diverge for two values of the angular momentum.

For $n = 3$, the situation is more complicated. Fig. 10.3 a) shows the fine structure for τ^- for a narrow band of initial angular momenta. In Fig. 10.3 b) we

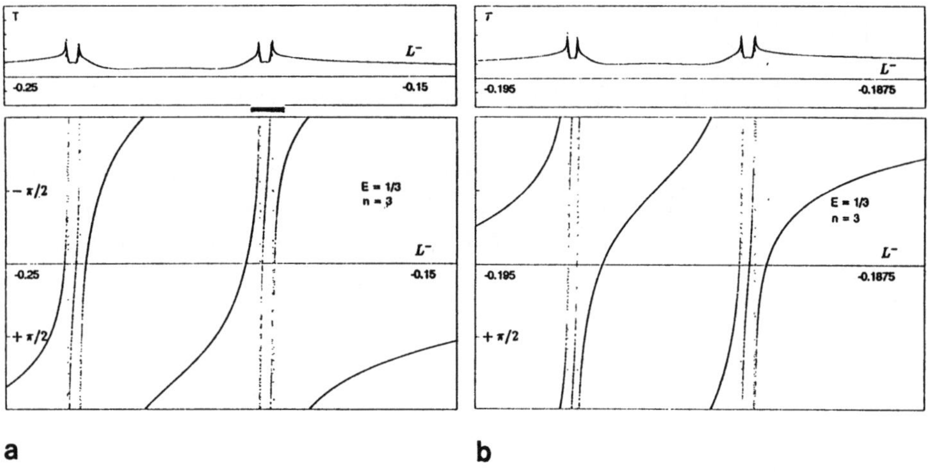

Fig. 10.3. a) A magnified view of Fig. 10.2 b); b) A magnified view of a)

show a narrow subregion of Fig. 10.3 a). The similarity of the last two figures is a sign of the Cantor structure of the set of initial angular momenta L^- for which $\tau(E, \varphi^-, L^-) = \infty$. The graphs below show the *deflection functions*, that is, the final angles φ^+ as functions of L^-. We shall come back to an interpretation of these graphs in Chap. 12.

Now we estimate the measure of the set of those scattering orbits of energy E whose time delay is larger than some positive time T. Therefore, we define

$$\kappa_E(T) := \int_{\mathbb{R}} \int_{S^1} \theta(\tau^\pm(E, \varphi, L) - T)\, d\varphi dL. \tag{10.3}$$

Remark 10.4 The measure $d\varphi\, dL$ used in Def. (10.3) is natural, as one sees from the following consideration. Outside the interaction zone, we may use the local canonical coordinates (r, φ, p_r, L) on P, with $q_1 = r\cos\varphi$, $q_2 = r\sin\varphi$ and $p_r := \vec{p} \cdot \vec{q} / |\vec{q}|$.

Then the symplectic two-form $\omega = dr \wedge dp_r + d\varphi \wedge dL$, so that the restriction of ω to a hypersurface $r \equiv R$ equals $d\varphi \wedge dL$.

In particular, the integral on the r.h.s. of (10.3) does not depend on the choice of incoming or outgoing coordinates, *i.e.* on the sign of $\tau^\pm$.

We may evaluate the asymptotic constants $\varphi^\pm(x)$, $L^\pm(x)$ of a scattering state $x \in s$ by taking the $R \to \infty$ limits of the φ, L coordinates of the two intersections between the orbit $\Phi^t(x)$ and the hypersurface $r \equiv R$.

In accordance with our general strategy, we estimate $\kappa_E(T)$ by going to the covering space.

Let the smooth surfaces $\mathbf{F}_E^\pm \subset \boldsymbol{\Sigma}_E$ be given by

$$\mathbf{F}_E^\pm := \left\{ (q,\dot{q}) \in \boldsymbol{\Sigma}_E \,\middle|\, |q| = R_{\text{vir}}, \pm\frac{d}{dt}\left|\boldsymbol{\eta}_E \boldsymbol{\Phi}_E^t(q,\dot{q})\right| |_{t=0} > 0 \right\}.$$

So every scattering geodesic flow line $\boldsymbol{\Phi}_E^t(x)$, $x \in \pi_E^{-1}(s \cap \Sigma_E)$ which meets the interaction zone $\mathbf{U}_E \subset \boldsymbol{\Sigma}_E$ enters it through $\mathbf{F}_E^-$ and leaves it through $\mathbf{F}_E^+$. For $x \in \mathbf{F}_E^\mp$, the exit time $\mathbf{T}_E^\pm(x)$ defined in (5.12) is the unique time for which $\boldsymbol{\Phi}_E^{\mathbf{T}_E^\pm(x)}(x) \in \mathbf{F}_E^\pm$.

The restriction of the symplectic two-form

$$\omega_E = \sum_{i,k} (\mathbf{g}_E)_{ik} dq_i \wedge d\dot{q}_k \quad \text{on } T\mathbf{M} \tag{10.4}$$

to $\mathbf{F}_E^\pm$ may be written as $\omega_E|_{\mathbf{F}_E^\pm} = dl \wedge d\cos\alpha$ where, similar to the coordinates on the Poincaré sections C_k ((6.7), (6.8)), l is a local parameter for the geodesic arc length on the curve $\boldsymbol{\eta}_E(\mathbf{F}_E^\pm) \subset \mathbf{M}$, and $\alpha(q,\dot{q})$ is the angle between $\dot{q}$ and the local direction of that (oriented) curve.

Lemma 10.5 *Let $n \geq 2$. Then there exist $C_9 > 0$, $C_7 > C_8 > 0$ such that for E large and $T > C_9$,*

$$\boldsymbol{\kappa}_E(T) := \int_{\mathbf{F}_E^-} \theta(\mathbf{T}_E^+(x) - T) dl \wedge d\cos\alpha \tag{10.5}$$

is bounded by

$$\exp(-C_7 \ln(E) T) < \boldsymbol{\kappa}_E(T) < \exp(-C_8 \ln(E) T). \tag{10.6}$$

Proof. We set $C_9 := t_{\max} \cdot (m_0 + 3)$, with $m_0 \in \mathbb{N}$ from Lemma 6.7.

The derivation of the upper bound is based on the following consideration. Let $x \in \mathbf{F}_E^-$ contribute to (10.5), that is, $\mathbf{T}_E^+(x) \geq T > C_9$. Then we know, using estimate (6.3), that the segment of the geodesic flow line $\boldsymbol{\Phi}_E^t(x)$ given by $0 < t < \mathbf{T}_E^+(x)$ intersects at least $m_1 + 1$ Poincaré surfaces C_k, with $m_1 := [T/t_{\max}] - 3$.

We denote by C_{k_0} the first Poincaré surface intersected by the flow line. Then the point of intersection is contained in the strip $V(k_0, \ldots, k_{m_1})$ with $(k_0, \ldots, k_{m_1})$ admissible in the sense of Chap. 6.

Therefore, we derive the upper bound in (10.6) by using the inequality

$$\boldsymbol{\kappa}_E(T) \leq \sum_{(k_0,\ldots,k_{m_1})} \int_{C_{k_0}} \chi(V(k_0, \ldots, k_{m_1})) dl \wedge d\cos\alpha, \tag{10.7}$$

where the summation is over all admissible sequences, and

$$\chi(V) : C_{k_0} \to \{0, 1\}$$

denotes the characteristic function of a set $V \subset C_{k_0}$. Now

$$\begin{aligned}
&\int_{C_{k_0}} \chi(V(k_0,\ldots,k_{m_1}))dl \wedge d\cos\alpha \\
&= \int_{I_{k_0}} (v_u(k_0,\ldots,k_{m_1})(i) - v_l(k_0,\ldots,k_{m_1})(i))\,di \\
&< C\cdot\exp(-J_l m_1 \ln(E)),
\end{aligned}$$

by Lemma 6.7 and the fact that the length $\mathcal{L}^E(\mathbf{d}_{k_0})$ of the interval I_{k_0} is uniformly bounded (above and below) for E large.

We have $2(n-1)\cdot(2n-3)^{m_1}$ admissible sequences $(k_0,\ldots,k_{m_1})$ in (10.7) so that

$$\boldsymbol{\kappa}_E(T) < 2(n-1)C\cdot\exp\left(m_1(\ln(2n-3) - J_l\ln(E))\right).$$

For E large, we may thus find a constant C_8 as appearing in the upper bound of (10.6).

To derive the lower bound, let $m_2 := [T/t_{\min}]$. For $(k_0,\ldots,k_{m_2})$ admissible, consider a point y in the set

$$V(k_0,\ldots,k_{m_2}) \cap (C_{k_0} \setminus \cup_{k_{-1}} W(k_{-1},k_0)).$$

y appears as the first intersection $y = \boldsymbol{\Phi}_E^t(x)$ at time t of the flow line through a unique point $x \in \mathbf{F}_E^-$ with a Poincaré section C_k, and $\mathbf{T}_E^+(x) \geq T$.

Thus, similar to (10.7), the lower bound is based on the inequality

$$\boldsymbol{\kappa}_E(T) \geq \sum_{(k_0,\ldots,k_{m_2})} \int_{C_{k_0}} \chi\left(V(k_0,\ldots,k_{m_2}) \cap (C_{k_0} \setminus \cup_{k_{-1}} W(k_{-1},k_0))\right) dl \wedge d\cos\alpha \tag{10.8}$$

where, again, the summation is over all admissible sequences and χ is the characteristic function.

Now we know from Lemma 6.6 that

$$v_u(k_0,k_1)(i) < 0 < w_l(k_{-1},k_0)(i) \quad \text{for } i \in I_{k_0} = \left[-\tfrac{1}{2}\mathcal{L}^E(\mathbf{d}_{k_0}), \tfrac{1}{2}\mathcal{L}^E(\mathbf{d}_{k_0})\right] \text{ large,} \tag{10.9}$$

(and conversely for i small), that is, the W strips separate from the V strips on the sides of the Poincaré section. As one concludes from a Euclidean estimate, one even finds an $\varepsilon > 0$ such that (10.9) holds for all $i \in \left[\frac{1}{2}\mathcal{L}^E(\mathbf{d}_{k_0}) - \varepsilon, \frac{1}{2}\mathcal{L}^E(\mathbf{d}_{k_0})\right]$ for E large.

Thus

$$\boldsymbol{\kappa}_E(T) > 2(n-1)\varepsilon\cdot\exp\left(m_2(\ln(2n-3) - J_u\ln(E))\right)$$

using (6.21), which proves the lower bound in (10.6). □

Theorem 10.6 *For all scattering states $x \in s$ the time delay $\tau(x)$ is bounded from below by $\tau(x) > -C(H(x))$, where the positive function $C(E)$ tends to zero as $E \to \infty$.*

Similarly, for $n = 1$, $\tau(x) < C(H(x))$.

For $n \geq 2$, there are constants $C_7 > C_8 > 0$ such that for E large and $T \geq 1$ the measure of the bounded orbits of energy E with time delay larger than T is bounded by

$$\exp(-C_7\sqrt{E}\ln(E) \cdot T) \leq \kappa_E(T) \leq \exp(-C_8\sqrt{E}\ln(E) \cdot T). \tag{10.10}$$

Proof. We estimate the time delay by setting $\tau(x) = \tau_1(x) + \tau_2(x)$, where

$$\tau_1(x) := \int_{\mathbb{R}} \sigma(R_{\mathrm{vir}}) \circ \Phi^t(x)dt$$

denotes the time spent within the interaction zone. Then there exists a uniform bound of the form

$$|\tau_2(x)| < \tfrac{1}{2}C(H(x)),$$

the reason being the following. The time spent by a Kepler orbit of energy E within a ball of radius R_{vir} centered at the origin is bounded from above by $2R_{\mathrm{vir}}/\sqrt{2E}$, see (2.16).

Therefore, up to a term going to zero as $E \to \infty$, $\tau_2(x)$ is the difference of the times spent by the orbit $\Phi^t(x)$ and the asymptotic orbits within the annulus of inner radius R_{vir} and outer radius R, as $R \to \infty$.

That difference may be uniformly bounded by some constant which goes to zero as $E \to \infty$ (use the integral equation (2.32) employed in the proof of Prop. 2.7).

Thus we have shown that $\tau(x) > -C(H(x))$. The upper bound for the case $n = 1$ follows easily, since we know that in that case for $H(x)$ large,

$$\int_{\mathbb{R}} \sigma(R_{\mathrm{vir}}) \circ \Phi^t(x)dt < \frac{CR_{\mathrm{vir}}}{\sqrt{2H(x)}}.$$

To derive the bounds (10.10) for the case $n \geq 2$, we estimate $\tau_1(x)$ using symbolic dynamics.

We know from Lemma 10.2 that for E large, all scattering states $x \in s$ of energy $H(x) = E$ and with time delay $\tau(x) \geq 1$ enter the ball of radius R_{vir} in the configuration plane. Let

$$F_E^{\pm} := \{(\vec{q}, \vec{p}) \in P \mid H(\vec{q}, \vec{p}) = E, |\vec{q}| = R_{\mathrm{vir}}, \pm\vec{q} \cdot \vec{p} > 0\}.$$

Then $F_E^{\pm}$ are two-dimensional submanifolds of Σ_E, and we want to estimate the unique return time $T_E^+(x)$, $x \in F_E^- \cap s$, for which we have $\Phi^{T_E^+(x)}(x) \in F_E^+$. Obviously, $T_E^+(x)$ equals $\tau_1(x)$, and we shall estimate $T_E^+(x)$ using the covering construction.

If $T \geq 1$, we may calculate $\kappa_E(T)$ using the formula

$$\kappa_E(T) = \int_{F_E^-} \theta(\tau(x) - T)\omega|_{F_E^-}, \tag{10.11}$$

since every energy E orbit with time delay ≥ 1 meets F_E^- exactly once, and since by Remark 10.4 the map $F_E^- \to S^1 \times \mathbb{R}$, $x \mapsto (\varphi^-(x), L^-(x))$ transforms the restriction $\omega|_{F_E^-}$ of the symplectic two-form to the measure $d\varphi \wedge dL$ which appears in (10.3).

Instead of (10.11), we will estimate

$$\kappa_E^1(T) = \int_{F_E^-} \theta(\tau_1(x) - T)\omega|_{F_E^-}. \tag{10.12}$$

As we shall see later, $\kappa_E(T)$ may be uniformly bounded by $\kappa_E^1(T)$ for $T \geq 1$, since the difference $\tau_2(x) = \tau(x) - \tau_1(x)$ is small.

The two-fold covering $\pi_E : \boldsymbol{\Sigma}_E \to \Sigma_E$ introduced in Prop. 3.1 restricts to a two-fold covering $\mathbf{F}_E^{\pm} \to F_E^{\pm}$, and

$$\pi_E^* \omega|_{\Sigma_E} = \sqrt{2E}\omega_E$$

by (3.4) and (10.4). So

$$\kappa_E^1(T) = \tfrac{1}{2}\sqrt{2E} \int_{\mathbf{F}_E^-} \theta(T_E^+(\pi_E(x)) - T) dl \wedge d\cos\alpha. \tag{10.13}$$

We can find constants $c' > 1 > c'' > 0$ such that for all large E and all $x \in \mathbf{F}_E^-$

$$c''\sqrt{2E}T_E^+(\pi_E(x)) < \mathbf{T}_E^+(x) < c'\sqrt{2E}T_E^+(\pi_E(x)), \tag{10.14}$$

since the conformal factor $\sqrt{1 - V(q)/E}$ in the Jacobi metric goes to one uniformly outside shrinking neighbourhoods of the singularities.

Inserting (10.14) into (10.13) we get

$$\sqrt{E/2}\,\boldsymbol{\kappa}_E(c'\sqrt{2E}T) < \kappa_E^1(T) < \sqrt{E/2}\,\boldsymbol{\kappa}_E(c''\sqrt{2E}T).$$

We insert estimate (10.6) in that inequality. By changing constants, the effect of τ_2 can be absorbed. Thus we have proven (10.10). □

11. Measure of the Bound States

In this chapter we discuss the question of the Liouville measure $\lambda(b_+)$ of the set

$$b_+ := \{x \in b \subset P \mid H(x) > 0\}$$

of positive energy bound states and relate it to time delay.

We know from Thm. 6.11 that $\lambda_E(b_E) = 0$ for E large. On the other hand, for the case of the Yukawa potential $V(\vec{q}) = -\exp(-\mu\,|\vec{q}\,|)/\,|\vec{q}\,|$, $\lambda_E(b_E) > 0$ for $E < \mu e^{-g}/2g$, where $g = (1+\sqrt{5})/2$ is the golden mean. This follows readily from the form of the effective potential (*i.e.*, including the centrifugal barrier) which then exhibits positive local minima for suitable values of the angular momentum.

So already for $n = 1$ we cannot expect $\lambda(b_+) = 0$ in general.

In these cases the *threshold energy*

$$E_{\mathrm{th}} := \inf\left\{E \in \mathbb{R} \mid \lambda(b \cap H^{-1}([E,\infty))) = 0\right\} \tag{11.1}$$

is strictly larger than zero. Yet there are Coulombic potentials for which much of our previous analysis applies for *all* positive energies and $E_{\mathrm{th}} = 0$:

Theorem 11.1 *Let $V < 0$ be a Coulombic potential whose logarithm is superharmonic,* i.e. $\Delta \ln |V| \geq 0$.

Then the conclusions of Thm. 6.11 hold true for all $E > 0$. In particular, $E_{\mathrm{th}} = 0$ so that $\lambda(b_+) = 0$.

Proof. We have

$$\Delta \ln |V| = \frac{V\Delta V - (\nabla V)^2}{V^2} \geq 0.$$

Thus $\Delta V \leq 0$ which implies for all $E > 0$

$$K_E = E \cdot \frac{(E-V)\Delta V + (\nabla V)^2}{2(E-V)^3} \leq 0,$$

using eq. (3.16). In other words, ln-superharmonicity of V leads to negative Gaussian curvature $K_E \leq 0$ for all positive energies.

Moreover, given $E > 0$, there exists a time $\mathbf{T}_{\mathrm{vir}} > 0$ after which any geodesic segment either leaves the interaction zone $\mathbf{G}_E$ of radius $R_{\mathrm{vir}}(E)$ or has felt a nonvanishing force $\nabla V \neq 0$. But for ln-superharmonic potentials, a nonvanishing gradient implies a Laplacian $\Delta V < 0$. So geodesics with large escape time meet regions with strictly negative Gaussian curvature.

Therefore, it is very easy to find a strictly invariant cone field (S_u^E, S_l^E) on $\mathbf{U}_E$ in the sense of Def. 5.6. For example, we may choose $S_l^E \equiv 0$ and S_u^E equal to a constant which is larger than $\sqrt{-K_E}$ everywhere in the interaction zone. Then the Riccati type equation (5.11) leads immediately to the invariance property.

The starting point of symbolic dynamics is the existence of closed geodesics $\mathbf{c}_l$ and dual geodesic segments $\mathbf{d}_l$ which is guaranteed by Lemma 6.1 and 6.2 since $V_{\max} = 0$ by assumption. The statement of Lemma 6.3 holds true since focal points do not exist if $K_E \leq 0$.

The basic estimate (6.3) then follows from the observation that we may bound the length $\mathcal{L}^E(c)$ of the geodesic segments from above by the diameter of the (simply connected) regions considered in Lemma 6.4. The rest of the analysis is similar to the general high-energy case. □

Remarks 11.2

1. Convex combinations $V_t := tV_1 + (1-t)V_0$, $0 \leq t \leq 1$ of ln-superharmonic Coulombic potentials $V_0, V_1 < 0$ are ln-superharmonic:

$$\begin{aligned} &-V_t \Delta V_t + (\nabla V_t)^2 \\ &\quad = t^2(-V_1\Delta V_1 + (\nabla V_1)^2) + (1-t)^2(-V_0\Delta V_0 + (\nabla V_0)^2) \\ &\qquad + t(1-t)(-V_1\Delta V_0 - V_0\Delta V_1 + 2\nabla V_0 \nabla V_1) \\ &\quad \leq t(1-t)(-V_1\Delta V_0 - V_0\Delta V_1 + 2\nabla V_0\nabla V_1) \\ &\quad \leq -t(1-t)\left(\sqrt{V_1\Delta V_0} - \sqrt{V_0\Delta V_1}\right)^2 \leq 0. \end{aligned}$$

2. $\ln(|\vec{q}\,|)$ is harmonic so that the logarithm of a Kepler potential is harmonic: $\Delta \ln|-Z/|\vec{q}\,|| = 0$. Purely Coulombic potentials

$$V(\vec{q}) = -\sum_{l=1}^{n} \frac{Z_l}{|\vec{q} - \vec{s}_l|}$$

are strictly ln-superharmonic in the sense $\Delta \ln|V| > 0$ for $n \geq 2$ (one may directly check the strict inequality for $n = 2$ and apply convex combinations for $n > 2$). This means that for $n \geq 2$ and a smooth additional potential W which decays fast, $V_a(\vec{q}) := -\sum_{l=1}^{n} \frac{Z_l}{|\vec{q}-\vec{s}_l|} + aW(\vec{q})$ is ln-superharmonic for $|a|$ small.

3. The motion Φ^t in these potentials is *integrable* on the positive energy submanifold $P_+ := \{x \in P \mid H(x) > 0\}$ of phase space P, in the following sense.

On the open submanifold $s \subset P_+$ of scattering states we consider the function L^-, that is, the initial asymptotic angular momentum. If V meets eqs. (2.35), then $L^-|_s$ is smooth. By definition it commutes with H, *i.e.* $\{L^-, H\} = 0$. Moreover, L^- is an independent constant of motion , that is, $dL^- \wedge dH$ does not vanish on s, since we may vary the initial angular momentum keeping the energy fixed.

But by Thm. 11.1 the measure $\lambda(P_+ \setminus s) = 0$. So we are justified to speak of integrability, observing that standard definitions of integrability (like the one by Abraham and Marsden [1], Def. 5.2.20) allow for linear dependence of constants of motion on measure-zero sets. In Knauf [29], the reader finds a discussion of why such a definition of integrability is natural for motion in a potential.

On the other hand, it has been shown by Bolotin [5] that for the purely coulombic case ($W = 0$), $n > 2$ centres and $E > 0$ there does not exist an *analytic* constant of the motion which is non-constant on Σ_E. The interested reader finds a discussion of this and related theorems in Fomenko [12].

If our potential V is not ln-superharmonic, then the low-energy dynamics will depend very much on the specific form of V. Still, we have the following general result:

Theorem 11.3 *Let V be a Coulombic potential with asymptotic charge Z_∞.*

Then if $Z_\infty \neq 0$, the measure $\lambda(b_+)$ of the positive energy bounded orbits is finite, and the virial radius $R_{\rm vir}$ may be choosen to be energy-independent.

If $Z_\infty = 0$ and $|\nabla V(\vec{q})| < C\,|\vec{q}|^{-3-\epsilon}$ for some $C, \epsilon > 0$ and $|\vec{q}| \geq R_{\rm min}$, then $\lambda(b_+) < \infty$, too.

Proof. We have $\lambda(b_+) = \int_0^\infty \lambda_E(b_E)dE = \int_0^{E_{\rm th}} \lambda_E(b_E)dE$ for the threshold energy $E_{\rm th}$ defined in (11.1). Furthermore, $\lambda_E(b_E) < \infty$ for each $E > 0$ since the bounded orbits stay inside the region of radius $R_{\rm vir}(E)$. So the only possible source for an infinite measure $\lambda(b_+)$ is a divergence of $\lambda_E(b_E)$ as $E \searrow 0$.

We treat the case $Z_\infty > 0$ first and show the existence of an energy-independent virial radius $R_{\rm vir}$ such that the r.h.s. $2(E - V(\vec{q})) - \vec{q}\cdot\nabla V(\vec{q})$ of the virial identity (2.23) is greater than zero for $E \geq 0$ and $|\vec{q}| \geq R_{\rm vir}$. The asymptotic property (2.4) of Coulombic potentials together with our assumption $Z_\infty > 0$ imply that $V(\vec{q}) < 0$ for $|q|$ large, since we know that $\delta V(\vec{q}) := V(\vec{q}) - (-Z_\infty/\,|\vec{q}|)$ is bounded by $|\delta V(\vec{q})| < |\vec{q}|^{-1-\epsilon}/(1+\epsilon)$.

Thus

$$\begin{aligned} -2V(\vec{q}) - \vec{q}\cdot\nabla V(\vec{q}) &\geq \frac{2Z_\infty}{|\vec{q}|} - \frac{2}{1+\epsilon}|\vec{q}|^{-1-\epsilon} - \frac{Z_\infty}{|\vec{q}|} - \frac{1}{1+\epsilon}|\vec{q}|^{-1-\epsilon} \\ &> (Z_\infty - 3\,|\vec{q}|^{-\epsilon})/\,|\vec{q}| \geq 0 \end{aligned}$$

for $|\vec{q}| \geq R_{\rm vir} := \max((3/Z_\infty)^{1/\epsilon}, R_{\rm min})$.

In the case $Z_\infty < 0$ we may choose $R_{\rm vir} := \max(Z_\infty^{-1/\epsilon}, R_{\rm min})$. Then for $|\vec{q}| \geq R_{\rm vir}$ the r.h.s. of the virial inequality

$$2(E - V(\vec{q})) - \vec{q} \cdot \nabla V(\vec{q}) \geq -\vec{q} \cdot \nabla V(\vec{q}) > \frac{|Z_\infty|}{|\vec{q}|} - |\vec{q}|^{-1-\epsilon} \geq 0.$$

The case $Z_\infty = 0$ must be treated differently.

Since $|\nabla V(\vec{q})| < C\,|\vec{q}|^{-3-\epsilon}$ for $|\vec{q}| \geq R_{\text{min}}$, the r.h.s. of the virial identity can be estimated by

$$\vec{p}^{\,2} - \vec{q} \cdot \nabla V(\vec{q}) \geq \vec{p}^{\,2} - \frac{C}{|\vec{q}|^{2+\epsilon}} \text{ for } |\vec{q}| \geq R_{\text{min}}$$

so that the bound states in the region $|\vec{q}| \geq R_{\text{min}}$ have a measure

$$\begin{aligned} &\lambda\left(b_+ \cap \{(\vec{q},\vec{p}) \in P \mid |\vec{q}| \geq R_{\text{min}}\}\right) \\ \leq\ & \int_{|\vec{q}| \geq R_{\text{min}}} d^2q \int_{\mathbb{R}^2} d^2p\,\theta(C\,|\vec{q}|^{-2-\epsilon} - |\vec{p}|^2) \\ =\ & 2\pi^2 \int_{R_{\text{min}}}^{\infty} \frac{C}{q^{2+\epsilon}} q\,dq = 2C\pi^2 R_{\text{min}}^{-\epsilon}/\epsilon < \infty. \end{aligned}$$

The measure of the bounded *inside* the region of radius R_{min} is clearly finite. So the assertion follows. □

Remark 11.4 In the asymptotically neutral ($Z_\infty = 0$) case, an estimate of the form $|\nabla V(\vec{q})| < C \cdot |\vec{q}|^{-3}$ for $|\vec{q}| \geq R_{\text{min}}$ is not sufficient to guarantee that $\lambda(b_+) < \infty$.

Consider, for example, a potential V which for $|\vec{q}| \geq R_{\text{min}}$ has the form $V(\vec{q}) = \cos(|\vec{q}|)/|\vec{q}|^2$.

In that case the measure $\lambda\left(b_+ \cap \{(\vec{q},\vec{p}) \in P \mid R_{\text{min}} \leq |\vec{q}| \leq R\}\right)$ diverges logarithmically as $R \to \infty$.

Now we derive a Levinson type theorem which relates the measure of the positive energy bound states to an integral over the time delay.

In [35], Narnhofer and Thirring derived such classical theorems for short-range potentials. For two space dimensions, their theorem relates the measure of *all* bound states to the sum of the integrated potential and the integrated time delay.

Conceptionally, Levinson type theorems compare quantities of the motion under consideration with quantities of a 'free' evolution which in our case is generated by the Kepler Hamiltonian H_∞. We are not primarily interested in the negative energy (bound) states, whose measure is infinite for $Z_\infty > 0$. Therefore, we derive a form of the theorem which relates quantities in the positive energy part of phase space.

Theorem 11.5 *Let V be a Coulombic potential with asymptotic charge $Z_\infty \geq 0$ which meets the estimate*

$$\left|\nabla V(\vec{q}) - Z_\infty \frac{\vec{q}}{|\vec{q}|^3}\right| < C\,|\vec{q}|^{-3-\epsilon} \quad \textit{for } |\vec{q}| \geq R_{\text{min}}. \tag{11.2}$$

Then

$$\lambda(b_+) = -2\pi \int_{\mathbb{R}^2} V_+(\vec{q})d^2q - \int \tau^-(E,\varphi,L)dE\,d\varphi\,dL, \tag{11.3}$$

with $V_+ := \max(V, 0)$.

In particular, the integrated time delay is zero if $V < 0$ *has a superharmonic logarithm,* i.e. $\Delta \ln V \geq 0$.

Proof. Let $\chi_b, \chi_s : P \to \{0,1\}$ be the characteristic functions of the bound states $b \subset P$ and the scattering states $s \subset P$, respectively.

Since we want to estimate quantities on compact phase space regions, we introduce for $a > 1$ the characteristic functions $\chi_a : P \to \{0,1\}$, $\chi_a^\infty : P_{\infty,+} \to \{0,1\}$, given by

$$\chi_a(x) := \begin{cases} 1 & \text{for } 1/a \leq H(x) \leq a \\ 0 & \text{otherwise} \end{cases}$$

and similarly for χ_a^∞, using the Hamiltonian function H_∞. Furthermore, we use the radial cutoffs $\sigma(R)$, $\sigma_\infty(R)$ introduced in Def. 10.1.

By asymptotic completeness of the flow Φ^t (Cor. 2.8) we have

$$\int_P \chi_a \cdot \sigma(R) d\lambda = \int_P \chi_a \cdot \sigma(R)(\chi_b + \chi_s) d\lambda. \tag{11.4}$$

From now on we assume $a > V_{\max}$. Then

$$\begin{aligned} \int_P \chi_a \cdot \sigma(R) d\lambda &= 2\pi \int_{|\vec{q}|\leq R} (a - \max(V(\vec{q}), 1/a)) d^2q \\ &= 2\pi \int_{|\vec{q}|\leq R} ((a - 1/a) - \max(V(\vec{q}) - 1/a, 0)) d^2q \\ &= \int_{P_{\infty,+}} \chi_a^\infty \cdot \sigma_\infty(R) d\lambda - 2\pi \int_{|\vec{q}|\leq R} \max(V(\vec{q}) - 1/a, 0) d^2q. \end{aligned} \tag{11.5}$$

We know from Prop. 2.7 that the Møller transformation $\Omega^+ : P_{\infty,+} \to s^+$ is measure-preserving and surjective. We need only consider the scattering states s, since $\lambda(s^+ \setminus s) = 0$ by asymptotic completeness. The first term in (11.5) transforms into

$$\begin{aligned} \int_{P_{\infty,+}} \chi_a^\infty \cdot \sigma_\infty(R) d\lambda &= \int_s (\chi_a^\infty \circ \Omega_*^+) \cdot (\sigma_\infty(R) \circ \Omega_*^+) d\lambda \\ &= \int_s \chi_a \cdot (\sigma_\infty(R) \circ \Omega_*^+) d\lambda, \end{aligned} \tag{11.6}$$

using energy conservation $H \circ \Omega^+ = H_\infty$.

Combining eqs. (11.4), (11.5) and (11.6) and observing that by Thm. 11.3

$$\lambda(b_+) = \lim_{a\to\infty} \lim_{R\to\infty} \int_P \chi_a \cdot \sigma(R) \cdot \chi_b \, d\lambda < \infty,$$

we arrive at the equation

$$\lambda(b_+) = -2\pi \int_{\mathbb{R}^2} V_+(\vec{q}) d^2q + \lim_{a\to\infty} \lim_{R\to\infty} \int_s \chi_a \cdot (\sigma_\infty(R) \circ \Omega_*^+ - \sigma(R)) d\lambda. \tag{11.7}$$

Now we show that the last term in (11.7) equals the integrated time delay. To this end we introduce the functions

$$f^{\pm}(a,R) := \int_s \chi_a \cdot \sigma^{\pm} \cdot (\sigma_\infty(R) \circ \Omega_*^{\pm} - \sigma(R))d\lambda \tag{11.8}$$

with $\sigma^{\pm}((\vec{q},\vec{p})) = \theta(\pm\vec{q}\cdot\vec{p})$.

We shall show that $\lim_{R\to\infty} f^{\pm}(a,R) = 0$. By time inversion symmetry it suffices to consider $f^{+}(a,R)$. First we note that $f^{+}(a,R) \leq f_1(a,R) - f_2(a,R)$ for the positive functions

$$f_1(a,R) := \int_s \chi_a \cdot \sigma^{+} \cdot (\sigma_\infty(R) \circ \Omega_*^{\pm}) \cdot (1-\sigma(R))d\lambda \tag{11.9}$$

and

$$f_2(a,R) := \int_s \chi_a \cdot \sigma^{+} \cdot (1-\sigma_\infty(R) \circ \Omega_*^{\pm}) \cdot \sigma(R)d\lambda, \tag{11.10}$$

so that $|f^{+}(a,R)| \leq f_1(a,R) + f_2(a,R)$.

To estimate $f_1(a,R)$, we must control the difference $|\vec{q}_0| - |\eta\Omega_*^{+}(\vec{q}_0,\vec{p}_0)|$ of the radii for outgoing initial conditions $(\vec{q}_0,\vec{p}_0)$, $\vec{q}_0\cdot\vec{p}_0 \geq 0$ of distance $\vec{q}_0 \geq R$ from the origin.

From the very definition of the Møller transformation we may conclude that this distance goes to zero as $R\to\infty$. But in order to bound the integral in (11.9), we must know the *rate* of convergence.

Therefore, similar to the proof of Prop. 2.7 we study the integral equation

$$(\tilde{\mathcal{F}}\vec{u})(t) := \int_t^\infty ds \int_s^\infty d\tau \left(\nabla V(\vec{q}(\tau)) - Z_\infty \frac{\vec{q}(\tau)+\vec{u}(\tau)}{|\vec{q}(\tau)+\vec{u}(\tau)|^3}\right) \tag{11.11}$$

with $(\vec{q}(t),\vec{p}(t)) := \Phi^t(\vec{q}_0,\vec{p}_0)$ for $1/a \leq H(\vec{q}_0,\vec{p}_0) \leq a$, $|\vec{q}_0| \geq R$ and $\vec{q}_0\cdot\vec{p}_0 \geq 0$.

We know that for R large, $\tilde{\mathcal{F}}$ is a contraction on

$$\mathcal{C}_0 = \left\{\vec{u} \in C([0,\infty), \mathbb{R}^2) \,\middle|\, \sup_t |\vec{u}(t)| < 1\right\}.$$

Moreover, for R large, we deduce from the virial identity (2.23) the simple inequality

$$|\vec{q}(t)| \geq \max\left(\sqrt{1/a}\, t, R\right) \tag{11.12}$$

which implies for R large

$$|(\tilde{\mathcal{F}}\vec{u})(0)| < 2C \int_0^\infty ds \int_s^\infty d\tau (\max(\tau/\sqrt{a}, R))^{-3-\epsilon},$$

using (11.2) and $|\vec{u}(\tau)| < 1$. So

$$\begin{aligned} |(\tilde{\mathcal{F}}\vec{u})(0)| &< 4Ca^{(3+\epsilon)/2} \int_0^\infty ds (\max(s, \sqrt{a}\cdot R))^{-2-\epsilon}, \\ &< 8CaR^{-1-\epsilon}. \end{aligned}$$

Thus only initial conditions $(\vec{q}_0, \vec{p}_0)$ with $R \le |\vec{q}| \le R + 8CaR^{-1-\epsilon}$ can contribute to the phase space integral (11.9) so that for R large, $f_1(a,R) < 8C(2\pi)^2a^2R^{-\epsilon}$. This implies

$$\lim_{R\to\infty} f_1(a,R) = 0. \tag{11.13}$$

To estimate $f_2(a,R)$ defined in (11.10), we split the phase space integral into two parts, multiplying the integrand by $1 = \sigma(R_{\mathrm{vir}}(1/a)) + (1 - \sigma(R_{\mathrm{vir}}(1/a))$.

The first integral goes to zero as $R \to \infty$ since we integrate over a compact set of initial conditions which only contribute if they are mapped by Ω_*^+ to points with radial distance $\ge R$ from the origin.

Initial conditions $(\vec{q}_0, \vec{p}_0)$ in the support of $\chi_a \cdot \sigma^+ \cdot (1 - \sigma(R_{\mathrm{vir}}(1/a))$ have a uniform upper bound C' for the distance $|\vec{q}_0 - \eta\Omega_*^+(\vec{q}_0, \vec{p}_0)|$, since they are outgoing. Only those initial conditions with $|\eta\Omega_*^+(\vec{q}_0, \vec{p}_0)| \ge R$ contribute to $f_2(a,R)$. Thus for R large we have $|\vec{q}_0| \ge R - C' > R/2$, and we apply the same reasoning as for $f_1(a,R)$ to show that

$$\lim_{R\to\infty} f_2(a,R) = 0. \tag{11.14}$$

Thus, using eqs. (11.7) to (11.14), we have shown that

$$\lambda(b_+) = -2\pi \int_{\mathbb{R}^2} V_+(\vec{q})d^2q + \lim_{a\to\infty}\lim_{R\to\infty} \int_s \chi_a \cdot \sigma^- \cdot (\sigma_\infty(R) \circ \Omega_*^+ - \sigma_\infty(R) \circ \Omega_*^-)d\lambda. \tag{11.15}$$

To relate the second term in (11.15) to time delay, we are allowed to confine integration to those initial conditions $(\vec{q}_0, \vec{p}_0)$ whose orbit $(\vec{q}(s), \vec{p}(s)) = \Phi^s(\vec{q}_0, \vec{p}_0)$ enters the ball of radius R at some time t which is then uniquely determined (the contribution of the other initial conditions vanishes as $R \to \infty$).

We write $x_0 := (\vec{q}_0, \vec{p}_0) = \Phi^{-t}(\vec{q}(t), \vec{p}(t))$ with $|\vec{q}(t)| = R$ and $\vec{q}(t)\cdot\vec{p}(t) \le 0$.

We can parametrize x_0 by the energy $H(x_0)$, the values $L^-(x_0)$, $\varphi^-(x_0)$ of the ingoing angular momentum and angle, and by the time $t(x_0)$.

We rewrite the Liouville measure $d\lambda$ using these canonical coordinates. Substituting $d\lambda = dt\, dH\, d\varphi^-\, dL^-$ in the second term of (11.15), we obtain

$$\int \left(\int_{\mathbb{R}} \chi_a \cdot \sigma^-(\Phi^t(x_0)) \cdot (\sigma_\infty(R) \circ \Omega_*^+ - \sigma_\infty(R) \circ \Omega_*^-)(\Phi^t(x_0))dt \right) dH\, d\varphi^- dL^-$$

with $(\vec{q}_0, \vec{p}_0) = x_0 = x_0(H, L^-, \varphi^-)$ with $|\vec{q}_0| = R$, $\vec{q}_0 \cdot \vec{p}_0 \le 0$.

We compare this expression to the one for time delay derived in Lemma 10.2. The two terms in (10.2) are interchanged by time inversion $(\vec{q}, \vec{p}) \mapsto (\vec{q}, -\vec{p})$. Thus both contribute the same term to the phase space integral:

$$\begin{aligned}\lim_{R\to\infty} \int\int_{\mathbb{R}} \chi_a \sigma^-(\Phi^t(x_0))(\sigma_\infty(R) \circ \Omega_*^+ - \sigma_\infty(R) \circ \Omega_*^-)(\Phi^t(x_0))dt\, dH\, d\varphi^- dL^- \\ = \int \chi_a \cdot \tau(x_0) dH\, d\varphi^-\, dL^-.\end{aligned}$$

This proves (11.3).

The integrated time delay is zero for the ln-superharmonic case since there $V_+ \equiv 0$ and by Thm. 11.1 the measure $\lambda(b_+) = 0$. □

12. The Differential Cross Section

The scattering transformation $S = \Omega_*^+ \circ \Omega^-$ contains complete information on the scattering process. As we have seen in the previous chapters, it exhibits many aspects of irregularity if $n \geq 3$. Nevertheless, the scattering transformation is not directly accessible in a (classical) scattering experiment.

Firstly, one typically cannot fix the initial angular momentum of the test particle. Secondly, it is hard to measure time delay, in particular in a classical setting where no interference effects exist.

What *is* accessible is the differential cross section $\frac{d\sigma}{d\theta+}(E,\theta^-,\theta^+)$. Informally speaking, we prepare incoming particles of energy E, initial angle θ^- and a uniform distribution of initial angular momenta L^-. Then $\frac{d\sigma}{d\theta+}(E,\theta^-,\theta^+)$ equals the number of particles scattered within a unit of time into the direction $[\theta^+,\theta^+ + d\theta^+]$, divided by the flux of the incoming particles with impact parameters $[b, b + db]$. We shall give a more precise definition below.

One could expect to see some trace of irregularity in the differential cross section, and in fact for all systems considered up to now numerical calculations of the cross section indicated the existence of so-called rainbow singularities on a Cantor set of angles.

The situation is different in our case. In Fig. 12.1 we show the differential cross section as a function of the final angle for the case of a sum of $n = 1, 2$ and 3 purely Coulombic potentials already considered in Chap. 10.

For $n = 1$ we obtain the two-dimensional Rutherford cross section

$$\frac{d\sigma}{d\theta+}(E,\theta^-,\theta^+) = \frac{Z}{4E}\frac{1}{\sin^2((\theta+ - \theta^-)/2)}. \tag{12.1}$$

It is remarkable that the differential cross sections for the cases $n = 2$ and 3 are very similar to the Rutherford cross section, except for a slight reduction of backscattering. So the complicated structure of the time delay and the deflection function (Figures 10.2 - 10.3) is not reflected in the cross section.

The reason for that discrepancy is, roughly speaking, the following. The deflection functions depicted in Figures 10.2 - 10.3 are strictly monotonic w.r.t. the initial angular momentum. It is clear from the definition of $\frac{d\sigma}{d\theta+}$ that extrema of the deflection function lead to singularities in the differential cross section. Since there are no extrema (except for the degenerate situation $L^- \to \pm\infty$), we have a nonsingular $\frac{d\sigma}{d\theta+}$ (except for the forward direction).

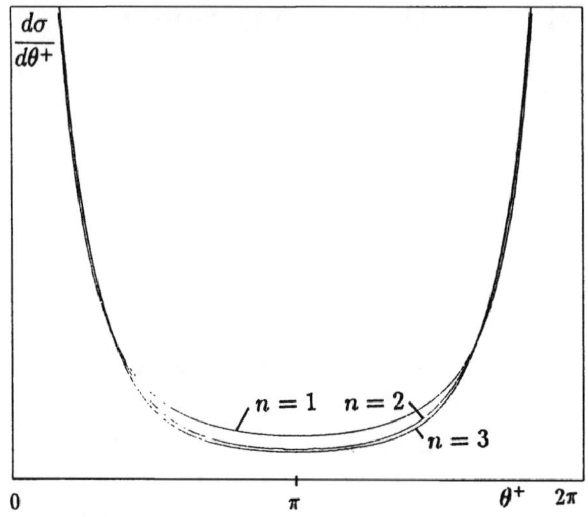

Fig. 12.1. The differential cross section $\frac{d\sigma}{d\theta^+}(E,\theta^-,\theta^+)$ as a function of θ^+ for $n = 1$, 2 and 3

As we shall see in Thm. 12.1 below, the monotonicity of the deflection function is caused by the negativity of the Gaussian curvature $K_E \leq 0$ for the cases under consideration.

In these cases we find a complete topological classification of the scattering orbits in Thm. 12.1 which serves as an input for the demonstration of smoothness of $\frac{d\sigma}{d\theta^+}$ in Thm. 12.3.

Similar to the classification of the closed geodesics in Thm. 4.6, the homotopy group $\pi_1(\mathbf{M})$ plays a central rôle. More precisely, a semidirect product $\pi_1(\mathbf{M}) \star \mathbb{Z}_2$ will classify the scattering orbits.

To introduce this product, we take for simplicity the based fundamental group $\pi_1(\mathbf{M}, \mathbf{s}_n)$. In Lemma 6.1 we introduced a set of closed geodesics $\mathbf{c}_1, \ldots, \mathbf{c}_{n-1} : \mathbb{R}/\mathbb{Z} \to \mathbf{G}_E$, starting from $\mathbf{c}_l(0) = \mathbf{s}_n$ and intersecting only at $\mathbf{s}_n$ which served as generators of $\pi_1(\mathbf{M}, \mathbf{s}_n)$. These closed geodesics are inverted by the covering transformation $G : \mathbf{M} \to \mathbf{M}$ defined in (3.3). So the induced automorphism

$$G^* : \pi_1(\mathbf{M}, \mathbf{s}_n) \to \pi_1(\mathbf{M}, \mathbf{s}_n)$$

given by

$$G^*([c]) := [c \circ G] \qquad \text{for a loop } c : S^1 \to \mathbf{M} \text{ based at } \mathbf{s}_n$$

is a non-trivial involution if $n > 1$. The multiplication

$$([c_1], m_1) \cdot ([c_2], m_2) := ([c_1] * m_1^*([c_2]), m_1 + m_2)$$

for $([c_i], m_i) \in \pi_1(\mathbf{M}, \mathbf{s}_n) \times \mathbb{Z}_2$ (with $\mathbb{Z}_2 = \{\mathrm{Id}, G\}$) fixes a semidirect product

$$\pi_1(\mathbf{M}, \mathbf{s}_n) \star \mathbb{Z}_2. \tag{12.2}$$

Obviously for $n = 1$, this group is simply $\mathbb{Z}_2$, whereas already for $n = 2$ we have a nonabelian group, namely $\mathbb{Z} \star \mathbb{Z}_2$. For $n \geq 3$, too, the product in (12.2) is not isomorphic to the direct product $\pi_1(\mathbf{M}, \mathbf{s}_n) \times \mathbb{Z}_2$.

Before stating our theorems, we shall recall a mathematically correct definition of cross section. In the physics literature the cross section is sometimes introduced as a function, whereas it really is a measure. The difference is of some importance because in general that cross section measure is not absolutely continuous w.r.t. Lebesgue measure. In our context, we shall *show* that under certain conditions the cross section measure *is* absolutely continuous if one excludes the forward direction, and that the Radon-Nikodym derivative, *i.e.* the differential cross section, is smooth.

We know from asymptotic completeness that the set of points

$$(E, L, \varphi) \in A^{\pm} := \mathbb{R}^+ \times \mathbb{R} \times S^1$$

which do not occur as asymptotic data $(E, L, \varphi) = (H(x), L^{\pm}(x), \varphi^{\pm}(x))$ for some scattering state $x \in s$ is of measure zero.

Thus the map $(H(x), L^-(x), \varphi^-(x)) \to (H(x), L^+(x), \varphi^+(x))$ induces a measurable map

$$(\bar{H}, \bar{L}, \bar{\varphi}) : A^- \to A^+.$$

For almost all $(E, \theta^-) \in \mathbb{R}^+ \times S^1$ the map

$$\varphi_{E,\theta^-} : \mathbb{R} \to S^1, \quad \varphi_{E,\theta^-}(L^-) := \bar{\varphi}(E, L^-, \theta^-) \tag{12.3}$$

is measurable.

We define the cross section measure $\sigma(E, \theta^-)$ on S^1 by

$$\sigma(E, \theta^-)(B) := \frac{1}{\sqrt{2E}} \cdot \lambda(\varphi_{E,\theta^-}^{-1}(B)) \tag{12.4}$$

for Borel sets $B \subset S^1$ and the Lebesgue measure λ on $\mathbb{R}$.

$\sigma(E, \theta^-)$ is not absolutely continuous w.r.t. Lebesgue measure because of a divergence in the forward scattering direction $\theta^+ = \theta^-$. So we remove that direction and define the *differential cross section* $\frac{d\sigma}{d\theta^+}(E, \theta^-, \theta^+)$ by

$$d\sigma(E, \theta^-) = \frac{d\sigma}{d\theta^+}(E, \theta^-, \theta^+) d\theta^+ \tag{12.5}$$

whenever $\sigma(E, \theta^-)$ is absolutely continuous w.r.t. Lebesgue measure on $S^1 \setminus \{\theta^-\}$.

Theorem 12.1 *Let V be a Coulombic potential with Gaussian curvature $K_{E_0} \leq 0$ for an energy E_0.*

Then, for $E > E_0$, in a neighbourhood of (θ_0^-, θ_0^+), $\theta_0^+ \neq \theta_0^-$, there is a natural bijection between the energy E orbits with asymptotic angles (θ^-, θ^+) and $(\pi_1(\mathbf{M}) \star \mathbb{Z}_2) \setminus \{\mathrm{Id}\}$. Moreover, the initial angular momenta $L_g^-(E, \theta^-, \theta^+)$ (which by definition verify the relation $\varphi_{E,\theta^-}(L_g^-(E, \theta^-, \theta^+)) = \theta^+$) are continuous in (θ^-, θ^+) and smooth if V meets the eqs. (2.35).

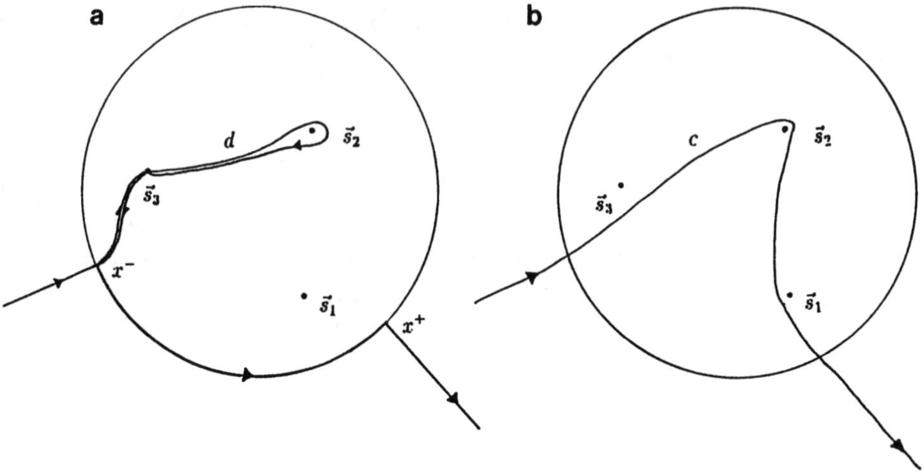

Fig. 12.2. a) The curve d which models b) the scattering orbit c.

The differential cross section (12.5) aquires the form

$$\frac{d\sigma}{d\theta^+}(E,\theta^-,\theta^+) = -\frac{1}{\sqrt{2E}} \sum_{g\in\left(\pi_1(\mathbf{M})\star\mathbb{Z}_2\right)\setminus\{\mathrm{Id}\}} \left(\frac{d\varphi_{E,\theta^-}}{dL}(L_g^-(E,\theta^-,\theta^+))\right)^{-1}, \tag{12.6}$$

where $d\varphi_{E,\theta^-}/dL(L_g^-) < 0$.

Remark 12.2 The basic idea of the proof of Thm. 12.1 is the following.

From the data (θ^-,θ^+) and $g = (\hat{g},m) \in \pi_1(\mathbf{M},\mathbf{s}_n) \star \mathbb{Z}_2$ we construct a regular *model curve* $d : \mathbb{R} \to \mathbf{M}$ which models a scattering geodesic, see Fig. 12.2 a). θ^- and θ^+ fix the asymptotic directions of the curve, the element $\hat{g} \in \pi_1(\mathbf{M})$ describes the sequence of loops of d inside the interaction zone, while $m \in \mathbb{Z}_2$ tells us whether d goes to infinity on the same sheet of the Riemann surface $\mathbf{M}$ on which it started.

Let $c : \mathbb{R} \to \mathbf{M}$ be a geodesic covering a scattering orbit with asymptotic angles (θ^-,θ^+), see Fig. 12.2 b). Then it turns out that for a suitable group element g we can homotope c to the model curve d, 'keeping the end points at infinity fixed'.

Proof. We begin with the definition of the model curve $d : \mathbb{R} \to \mathbf{M}$ for $g = (\hat{g},m)$, see Fig. 12.2 a). Let $x^\pm \in \mathbf{M}$ be points on the boundary $\partial\mathbf{G}_E$ projecting to

$$\pi(x^\pm) = \pm R_{\mathrm{vir}} \cdot \begin{pmatrix}\cos\theta^\pm \\ \sin\theta^\pm\end{pmatrix}.$$

There are two points x^- with the prescribed properties, and we choose one of them. The selection of the point x^+ will depend on the group element g.

The ingoing and outgoing segments of d have parameter values $|t| \geq 2$. We set $d(\pm 2) := x^{\pm}$ and let $d(t)$ project to the rays

$$\pi(d(\pm t)) = (|t| - 1) \cdot \pi(x^{\pm}), \quad |t| \geq 2$$

in the configuration plane. Assuming continuity of d, this assumption fixes the two segments.

We connect $d(-2) = x^-$ with the base point $d(-1) := \mathbf{s}_n$ of $\pi_1(\mathbf{M}, \mathbf{s}_n)$ by some regular curve $d(t)$, $-2 \leq t \leq -1$. We use that curve segment for all model curves with the same asymptotic angles (θ^-, θ^+).

The segment $d|_{[-1,0]}$ with end points $d(-1) = \mathbf{s}_n =: d(0)$ is chosen so that the homotopy class of $d|_{[-1,0]} \in \pi_1(\mathbf{M}, \mathbf{s}_n)$ equals $\hat{g}$.

If $m \in \mathbb{Z}_2$ is the identity, then we set $d(t) := d(-1-t)$ for $0 < t \leq 1$; if m equals the inversion G, then we set $d(t) := G(d(-1-t))$, that is, we take the mirror image of the path.

Finally, for $1 \leq t < 2$, we move along the boundary of the interaction zone by letting $d(t)$ project to

$$\pi(d(\pm t)) = R_{\text{vir}} \cdot \begin{pmatrix} \cos\left((t-1)\theta^+ + (2-t)(\theta^- + \pi)\right) \\ \sin\left((t-1)\theta^+ + (2-t)(\theta^- + \pi)\right) \end{pmatrix},$$

assuming that $|\theta^+ - (\theta^- + \pi)| < \pi$.

Then the choice of $x^+ = d(2)$ is fixed by the assumption of continuity of the curve d.

We say that two curves $d_0, d_1 : \mathbb{R} \to \mathbf{M}$ with $\lim_{t \to \pm\infty} |d_i(t)| = \infty$ are *scattering homotopic* if there exists a continuous map

$$H : \mathbb{R} \times [0,1] \to \mathbf{M}$$

with $H(t,0) = d_0(t)$ and $H(t,1) = d_1(f(t))$ for an increasing homeomorphism $f : \mathbb{R} \to \mathbb{R}$, and

$$\sup\{d(H(t,r), H(t,0)) \mid (t,r) \in \mathbb{R} \times [0,1]\} < \infty. \tag{12.7}$$

It is clear that two model curves for different scattering data (θ^-, θ^+), g are not scattering homotopic.

For $g \in (\pi_1(\mathbf{M}) \star \mathbb{Z}_2) \setminus \{\text{Id}\}$ we show the existence of a geodesic c scattering homotopic to d which will turn out to be unique up to reparametrization of time.

We shall construct c by shortening the curve segments $d_a := d|_{-a,a}$ for $a \geq 2$ keeping the end points fixed and then letting a go to infinity.

For each $a \geq 2$ we find a geodesic segment $c_a : [-a, a] \to \mathbf{M}$ with $c_a(\pm a) = d_a(\pm a)$ homotopic to d_a.

Since the Gaussian curvature $K_E \leq 0$, we know from Thm. 2.6.6 of [26] that c_a is unique.

Now we show that all these geodesic segments c_a enter a common ball of radius R, *i.e.*

$$\sup_{a \geq 2} \inf\{|c_a(t)| \mid -a \leq t \leq a\} < \infty. \tag{12.8}$$

Assume the contrary, *i.e.* the existence of a divergent sequence $\{a_i\}_{i \in \mathbb{N}}$ with

$$\inf\{|c_{a_i}(t)| \mid -a_i \leq t \leq a_i\} \geq i. \tag{12.9}$$

Then the integrated (absolute) geodesic curvature of the segments w.r.t. the Euclidean metric

$$\int_{-a_i}^{a_i} |k_g(c_{a_i})(t)| \, |\dot{c}_{a_i}(t)| \, dt = \int_{-a_i}^{a_i} \left| \frac{\langle J\dot{c}_{a_i}(t), \ddot{c}_{a_i}(t)\rangle}{|\dot{c}_{a_i}(t)|^3} \right| dt \tag{12.10}$$

goes to zero as $i \to \infty$, since we have the estimate

$$|c_{a_i}(t - t_i)| \geq \max(i, |t|/2) \qquad \text{for } i \text{ large}, \tag{12.11}$$

where $|c_{a_i}(t)|$ is minimal at the unique time t_i. Similar to the inequality (11.12), (12.11) follows from the virial identity (2.23). The integral in (12.10) of the geodesic curvature goes to zero since $\ddot{c}(t)$ is controlled by $|\nabla V|(\pi(c(t))$ which by (2.4) falls off like the inverse squared distance from the origin.

We know from Lemma 5.3 that

$$\frac{\mathcal{L}^\infty(c_{a_i})}{|c_{a_i}(a_i) - c_{a_i}(-a_i)|} \to 1 \tag{12.12}$$

so that c_{a_i} converges in a weak sense to a straight line connecting its end points.

Using (12.12) and assuming $\theta^- \neq \theta^+$, we shall show that it follows from our assumption (12.9) that the group element g used in the definition of the model curve d equals the identity Id.

First it is obvious from (12.12) that for i large there is a curve $\bar{c}_{a_i}$ with end points $\bar{c}_{a_i}(\pm a_i) = c_{a_i}(\pm a_i)$ which projects to a straight line $\pi(\bar{c}_{a_i})$, and that c_{a_i} is homotopic to $\bar{c}_{a_i}$.

Now it suffices to show the existence of a homotopy from the model curve $\hat{d}_{a_i}$ to $\bar{c}_{a_i}$, where $\hat{d} : \mathbb{R} \to \mathbf{M}$ is defined as d above except that we replace $g \in (\pi_1(\mathbf{M}, \mathbf{s}_n) \star \mathbb{Z}_2) \setminus \{\text{Id}\}$ by Id.

This is easily done since by the definition of $\hat{d}$, $\hat{d}(t) = \hat{d}(-1-t)$ for $0 \leq t \leq 1$, and since we may assume $\hat{d}(t) = \mathbf{s}_n$ for $-1 \leq t \leq 0$. Thus $\hat{d}_{a_i}$ is homotopic to a curve $\tilde{d}_{a_i} : [-a_i, a_i] \to \mathbf{M}$ with $\tilde{d}_{a_i}(t) := \hat{d}_{a_i}(t)$ for $t \leq -2$ and for $t \geq 1$, and $\tilde{d}_{a_i}(t) = x^-$ for $-2 \leq t \leq -1$. Pushing $\tilde{d}_{a_i}$ in the radial direction we see that it is homotopic to the straight line between its end points $c_{a_i}(\pm a_i)$ (this straight line (w.r.t. the Euclidean metric) is well-defined for a large since we assumed $\theta^- \neq \theta^+$!).

What we have proven up to now is estimate (12.8) for group elements $g \in (\pi_1(\mathbf{M}) \star \mathbb{Z}_2) \setminus \{\text{Id}\}$. Thus all the geodesic segments c_a, $a \geq 2$, enter a closed ball of common radius R.

By compactness of that ball a sequence of points $x_i := c_i(s_i)$ with radius $|c_i(s_i)| \leq R$ has an accumulation point $x \in \mathbf{M}$, and by going to a subsequence, if necessary, we not only have $x_i \to x$ but also convergence of $\dot{x}_i := \dot{c}_i(s_i)/|\dot{c}_i(s_i)|$ to a velocity $\dot{x}$. Now $(x, \dot{x}) \in \Sigma_E$ fix a geodesic

$$c : \mathbb{R} \to \mathbf{M} \qquad \text{with } (c(0), \dot{c}(0)) := (x, \dot{x})$$

which stays inside the interaction zone only a finite time (since the geodesic segments c_i minimize the length between their end points). In other words, $z := \pi_E((x, \dot{x})) \in \Sigma_E$ belongs to the set $s \subset P$ of scattering states.

We attribute to c_i the asymptotic angles

$$\theta_i^{\pm} := \varphi^{\pm}(z_i) \qquad \text{for } z_i := \pi_E((x_i, \dot{x}_i)) \in \Sigma_E.$$

Then the limits $\lim_{i\to\infty} \theta_i^{\pm}$ exist, since $p_1^{\pm} = \sqrt{2E}\cos\varphi^{\pm}$, $p_2^{\pm} = \sqrt{2E}\sin\varphi^{\pm}$, and $\vec{p}^{\,\pm} : s^{\pm} \to \mathbb{R}^2$ defined in (2.28) is continuous. We shall show that that these limits coincide with those of the model curve d, that is, that

$$\vec{p}^{\,\pm}(z) = \vec{P}^{\,\pm} \tag{12.13}$$

with $P_1^{\pm} := \sqrt{2E}\cos\theta^{\pm}$ and $P_2^{\pm} := \sqrt{2E}\sin\theta^{\pm}$.

Let $t_i^{\pm} \in \mathbb{R}$ be the times for which $(\vec{q}_i(t), \vec{p}_i(t)) := \Phi^t(z_i)$ meets the equation

$$\left(\vec{q}_i(t_i^{\pm}), \vec{p}_i(t_i^{\pm})\right) = \pi_E\left(c_i(\pm i), \dot{c}_i(\pm i)/\left|\dot{c}_i(\pm i)\right|\right),$$

so that they are the projections of the end points of the geodesic flow line segments.

Then

$$\begin{aligned} \left|\vec{p}^{\,\pm}(z) - \vec{P}^{\,\pm}\right| \;\leq\; & \lim_{i\to\infty}\left|\vec{p}^{\,\pm}(z) - \vec{p}^{\,\pm}(z_i)\right| + \lim_{i\to\infty}\left|\vec{p}^{\,\pm}(z_i) - \vec{p}_i(t_i^{\pm})\right| \\ & + \lim_{i\to\infty}\left|\vec{p}_i(t_i^{\pm}) - \sqrt{2E}\,\vec{q}_i(t_i^{\pm})/\left|\vec{q}_i(t_i^{\pm})\right|\right|. \end{aligned} \tag{12.14}$$

All three terms on the r.h.s. of (12.14) go to zero, implying (12.13). Continuity of $\vec{p}^{\,\pm}$ leads to the vanishing of the first term. The second part vanishes since

$$\vec{p}^{\,\pm}(z_i) - \vec{p}_i(t_i^{\pm}) = \int_{t_i^{\pm}}^{\infty} \nabla V(\vec{q}_i(t))dt.$$

The third term goes to zero, since the angular momentum $L(\Phi^t(z_i))$ is uniformly bounded in $t \in \mathbb{R}$ and $i \in \mathbb{N}$.

So we have shown that the asymptotic angles of the geodesic c are (θ^-, θ^+), and c is scattering homotopic to the model curve d.

Now we prove that the geodesic c is the *unique* geodesic which is scattering homotopic to the model curve d (modulo reparametrizations of time, of course).

Let $c_0, c_1 : \mathbb{R} \to \mathbf{M}$ be two geodesics which are scattering homotopic to d (and thus to each other). Then they are parallel at infinity in the following sense. Consider a shortest geodesic from $c_0(t)$ to the geodesic c_1. By (12.7) the lengths of these connecting geodesic segments are bounded uniformly in t, and for $|t|$ large they are unique.

But since the potential V decays at infinity, the geodesic curvature of geodesic segments far away from the origin goes to zero implying that the connecting geodesics intersect c_0 and c_1 with angles which both converge to $\pi/2$ as $|t| \to \infty$. Therefore, by the Gauss-Bonnet theorem, the lifts of c_0 and c_1 to the universal cover of $\mathbf{M}$ must bound a strip of integrated Gaussian curvature zero. Then either the curvature $K_E = 0$ everywhere inside the strip, or the two geodesics coincide, since we assumed $K_E \leq 0$. The first alternative cannot hold, for the following reason.

Since we assumed $\theta^+ \neq \theta^-$, $\vec{q}(t) := \pi(c_0(t))$ must meet regions in the configuration plane where the force $-\nabla V$ does not vanish. In those regions we have $\Delta V < 0$ by our assumption $K_{E_0} \leq 0$, which in turn implies $K_E(c_0(t)) < 0$.

After having demonstrated uniqueness, we show that for $\theta^+ \neq \theta^-$

$$\frac{d\varphi_{E,\theta^-}}{dL}(L^-) < 0 \qquad \text{if } \varphi_{E,\theta^-}(L^-) = \theta^+, \tag{12.15}$$

with φ_{E,θ^-} defined in (12.3).

So we have a scattering state $x \in s$ with

$$(H(x), L^-(x), \varphi^-(x)) = (E, L^-, \theta^-)$$

and $\varphi^+(x) = \theta^+$.

We must linearize the flow along the orbit $\Phi_E^t(x)$ through x and consider the one-parameter family of variations of the incoming state

$$(\delta E, \delta L^-, \delta\theta^-) = (0, \lambda, 0).$$

Alternatively, we may consider the following linearized equation on covering space $(\mathbf{M}, \mathbf{g}_E)$: Let $(c_0, \dot{c}_0) \in \boldsymbol{\Sigma}_E$ project to $x \in \Sigma_E$, *i.e.* $x = \pi_E((c_0, \dot{c}_0))$, and let $c : \mathbb{R} \to \mathbf{M}$ be the geodesic with initial conditions $(c_0, \dot{c}_0)$.

Instead of studying the linearized flow $T\Phi_E^t(x)$, we analyse the Jacobi equation

$$\ddot{Y}_\lambda(s) + K_E(c(s)) \cdot Y_\lambda(s) = 0 \tag{12.16}$$

with boundary conditions

$$\lim_{t \to -\infty} Y_\lambda(t) = -\lambda/\sqrt{2E}, \qquad \lim_{t \to -\infty} \dot{Y}_\lambda(t) = 0. \tag{12.17}$$

The eqs. (12.17) are the correct boundary conditions since for $t \to \pm\infty$ the geodesic $c(t)$ goes to spatial infinity where the Jacobi metric converges to the Euclidean metric. $\lambda/\sqrt{2E}$ then corresponds to the variation of the impact parameter $L^-/|\vec{p}^{\,-}|$.

The variation $\delta\varphi^+$ of the outgoing angle with λ equals $\lim_{t \to +\infty} \dot{Y}_\lambda(t)$ since $\dot{Y}_\lambda$ describes the variation of the velocity perpendicular to the geodesic $c(t)$ which is parametrized by arc length.

We claim that the limit $\lim_{t \to +\infty} \dot{Y}_\lambda(t)$ exists. First of all, $|K_E(\vec{q})| < C_1 |\vec{q}|^{-2-\epsilon}/E$ for $|\vec{q}| > R_{\min}$, using $\vec{q}$ coordinates, and the distance of $c(t)$ from the origin is larger than $|t|/2$ for $|t|$ large. Secondly, $Y_\lambda(t) = \mathcal{O}(t)$ for t large, as follows from a comparison argument. Then our claim follows by considering the expression

$$\dot{Y}_\lambda(t) - \dot{Y}_\lambda(0) = \int_0^t -K_E(c(s))Y_\lambda(s)ds.$$

Similarly, the boundary value problem (12.16), (12.17) has a unique solution,as we see by inspection of the integral equation

$$Y_\lambda(t) = -\int_{-\infty}^t \int_{-\infty}^s K_E(c(\tau)) \cdot Y_\lambda(\tau)d\tau\, ds + \frac{\lambda}{\sqrt{2E}}.$$

$|Y_\lambda(t)|$ as well as $\left|\dot{Y}_\lambda(t)\right|$ increases along the geodesic since

$$\dot{Y}_\lambda(t) = \int_{-\infty}^t K_E(c(s))Y_\lambda(s)ds$$

and $K_E \leq 0$. As we have seen, $K_E < 0$ somewhere on the geodesic, so that

$$\frac{d\varphi_{E,\theta^-}}{dL}(L^-) = \frac{\partial}{\partial\lambda} \lim_{t\to\infty} \dot{Y}_\lambda(t) < 0, \tag{12.18}$$

showing (12.15).

Thus for $E > E_0$ and $\theta^+ \neq \theta^-$ we can label the initial angular momenta in the set $\varphi_{E,\theta^-}^{-1}(\theta^+) \subset \mathbb{R}$ locally uniquely by $g \in (\pi_1(\mathbf{M}) \star \mathbb{Z}_2) \setminus \{\mathrm{Id}\}$.

(12.15) and an application of the implicit function theorem show that the initial angular momenta $L_g^-(E,\theta^-,\theta^+)$ meeting the relation

$$\varphi_{E,\theta^-}(L_g^-(E,\theta^-,\theta^+)) = \theta^+$$

are continuous in (θ^-,θ^+) and smooth if V meets the eqs. (2.35). The representation (12.6) of the differential cross section as a sum is then a direct consequence of our classification of the scattering orbits. □

Although we know that the single terms in (12.6) are smooth for $\theta^+ \neq \theta^-$ if (2.35) holds, we do not even know whether that sum converges. We will handle that question now.

Theorem 12.3 *Let V be a Coulombic potential meeting eqs. (2.35), with Gaussian curvature $K_{E_0} \leq 0$ for some energy $E_0 \geq V_{\max}$.*

Then for $E > E_0$ the differential cross section $\frac{d\sigma}{d\theta^+}(E,\theta^-,\theta^+)$ is a smooth function of the angles (θ^-,θ^+), for $\theta^+ \neq \theta^-$.

Proof. We show continuity in (θ^-,θ^+) for $\theta^+ \neq \theta^-$. The same kind of argument can be applied to show continuity of the derivatives.

We use eq. (12.6) to represent $\frac{d\sigma}{d\theta^+}$ as a sum of smooth functions. It is clear from the proof of Thm. (12.1) (see eq. (12.18)) that, except for a finite number of group elements $g \in (\pi_1(\mathbf{M}) \star \mathbb{Z}_2) \setminus \{\mathrm{Id}\}$,

$$-\frac{d\varphi_{E,\theta^-}}{dL}(L_g^-(E,\theta^-,\theta^+)) > C' > 0$$

is uniformly bounded from below even for $\theta^+ = \theta^-$. Moreover, except for a finite number of group elements, the initial angular momenta L_g^- are uniformly bounded in absolute value.

So there is a finite covering $\mathbb{T}^2 = \bigcup_{l=1}^k U_l$ of the two-torus $\mathbb{T}^2 = \{(\theta^-, \theta^+)\}$ of asymptotic angles by open sets, and there are k subsets

$$G_l \subset (\pi_1(\mathbf{M}) \star \mathbb{Z}_2) \setminus \{\mathrm{Id}\}$$

such that for $(\theta^-, \theta^+) \in U_l$

$$\frac{d\sigma}{d\theta^+}(E, \theta^-, \theta^+) = \Sigma_1 + \Sigma_2$$

where the first sum $\Sigma_1 := -(2E)^{-1/2} \cdot \sum_{g \notin G_l} (\frac{d\varphi_{E,\theta^-}}{dL}(L_g^-(E, \theta^-, \theta^+)))^{-1}$ contains only finitely many terms, and the second sum

$$\Sigma_2 := -(2E)^{-1/2} \cdot \sum_{g \in G_l} \left(\frac{d\varphi_{E,\theta^-}}{dL}(L_g^-(E, \theta^-, \theta^+)) \right)^{-1}$$

is integrable:

$$\int_{U_l} \Sigma_2(E, \theta^-, \theta^+) d\theta^- d\theta^+ \leq (2E)^{-1/2} \int_{S^1 \times \{|L^-| \leq L_{\max}\}} d\theta^- dL^- < \infty,$$

where

$$L_{\max} := \sup\{L_g^-(E, \theta^-, \theta^+) \mid g \in G_l, (\theta^-, \theta^+) \in U_l\} < \infty.$$

and we have used the change of variables

$$d\theta^- \wedge d\theta^+ = \frac{\partial \theta^+}{\partial L^-}(E, \theta^-, L_g^-(E, \theta^-, \theta^+)) d\theta^- \wedge dL^-.$$

We demonstrate that Σ_2 is continuous on U_l by proving the existence of a Lipschitz constant $C > 0$ such that for all $l \in \{1, \ldots, k\}$ and all scattering data $(\theta^-, \theta^+) \in U_l$, $g \in G_l$

$$\left| \frac{\partial}{\partial \theta^\pm} \left(\frac{d\varphi_{E,\theta^-}}{dL}(L_g^-) \right)^{-1} / \left(\frac{d\varphi_{E,\theta^-}}{dL}(L_g^-) \right)^{-1} \right| = \left| \frac{\partial}{\partial \theta^\pm} \left(\ln(\frac{d\varphi_{E,\theta^-}}{dL}(L_g^-)) \right) \right| \leq C. \tag{12.19}$$

We then enumerate the elements of G_l by g_k, $k \in \mathbb{N}$ and set $f_k(\theta) := \left(\frac{d\varphi_{E,\theta^-}}{dL}(L_{g_k}^-)(E, \theta^-, \theta^+)) \right)^{-1}$. From (12.19) one gets with $\theta := (\theta^-, \theta^+)$

$$\left| \frac{\partial}{\partial \theta^\pm} f_k(\theta) \right| < C f_k(\theta)$$

which implies

$$\left| \frac{\partial}{\partial \theta^\pm} \sum_{k=1}^N f_k(\theta) \right| < C \sum_{k=1}^N f_k(\theta).$$

Setting $g_N(\theta) := \ln \sum_{k=1}^{N} f_k(\theta)$ we thus have

$$\left|\frac{\partial}{\partial\theta^\pm} g_N(\theta)\right| \leq C. \tag{12.20}$$

The pointwise limit $g := \lim_{N\to\infty} g_N$ exists by monotonicity (it may be infinite) and by (12.20) satisfies

$$|g(\theta_1) - g(\theta_2)| \leq C\,|\theta_1 - \theta_2|\,. \tag{12.21}$$

Since $\sum_{k=1}^{\infty} f_k(\theta) > 0$ is integrable, g is finite at least at some fixed point θ. By (12.21) it is then finite everywhere and Lipschitz, which proves our claim that Σ_2 is (Lipschitz) continuous on U_l.

Thus is remains to prove (12.19). To this end we shall represent the l.h.s. of (12.19) as the time integral of the solution of a linear inhomogeneous differential equation of first order which arises from linearization of the flow along the orbit.

For the scattering data $(\theta^-, \theta^+) \in U_l$, $g \in G_l$ we consider the associated geodesics

$$c(\cdot) \equiv c_g(\cdot;\theta^-,\theta^+) : \mathbb{R} \to \mathbf{M}$$

parametrized by arc length and synchronized by demanding that

$$\lim_{t\to-\infty}\Big(|c_g(t;\theta^-,\theta^+)| - |c_g(t;\theta_0^-,\theta_0^+)|\Big) = 0 \text{ for all } (\theta^-,\theta^+) \in U_l.$$

This makes c_g smooth in all its arguments. Moreover, the variational vector fields

$$\vec{Y}_g^\pm(t;\theta^-,\theta^+) := \frac{\partial}{\partial\theta^\pm} c_g(t;\theta^-,\theta^+)$$

are perpendicular to the direction $\dot{c}(t)$ of the geodesic.

So $Y^\pm(t) \equiv Y_g^\pm(t;\theta^-,\theta^+) := \left\langle \vec{Y}_g^\pm(t;\theta^-,\theta^+), J\dot{c}_g(t;\theta^-,\theta^+)\right\rangle$ (with $J = \left(\begin{smallmatrix}0 & -1\\ 1 & 0\end{smallmatrix}\right)$) meets the Jacobi equation

$$\frac{\partial^2}{\partial t^2} Y^\pm(t) + K_E(c(t)) Y^\pm(t) = 0, \tag{12.22}$$

with the boundary values

$$\lim_{t\to\pm\infty} \frac{\partial}{\partial t} Y^\pm(t) = \pm 1, \quad \lim_{t\to\mp\infty} \frac{\partial}{\partial t} Y^\pm(t) = 0, \tag{12.23}$$

and

$$\lim_{t\to-\infty} Y^+(t) = -\frac{1}{\sqrt{2E}} \left(\frac{d\varphi_{E,\theta^-}}{dL}\right)^{-1}. \tag{12.24}$$

Furthermore, we know that $\pm Y^\pm(t) > 0$ and $\pm\frac{\partial}{\partial t} Y^\pm(t) \geq 0$, using (12.22), (12.23) and the negativity of K_E. So

$$S(t) \equiv S_g(t;\theta^-,\theta^+) := \frac{\partial}{\partial t} \ln Y_g^+(t;\theta^-,\theta^+) \tag{12.25}$$

is smooth, meets the Riccati equation

$$\dot{S}(t) + S^2(t) + K_E(c(t)) = 0, \tag{12.26}$$

and $0 \le S(t) \le |\min(K_E)|^{1/2}$.

Integrating (12.25) from $-\infty$ to t and using (12.24), we have

$$\ln\left|\sqrt{2E}\frac{d\varphi_{E,\theta^-}}{dL}(L_g^-(E,\theta^-,\theta^+))\right| = \int_{-\infty}^t S(\tau)d\tau + \ln S(t) - \ln \dot{Y}^+(t). \tag{12.27}$$

We remark that in view of (12.23) the integral on the r.h.s. of (12.27) diverges logarithmically as $t \to \infty$. We shall show, however, that its derivative w.r.t. $\theta^\pm$ converges. Thus we may differentiate (12.27) to obtain the estimate (12.19), *i.e.* the existence of a global Lipschitz constant C.

So let

$$Z^\pm(t) \equiv Z_g^\pm(t;\theta^-,\theta^+) := \frac{\partial}{\partial\theta^\pm} S_g(t;\theta^-,\theta^+).$$

By taking the derivative of (12.26), we see that the $Z^\pm(t)$ satisfy the equation

$$\frac{\partial}{\partial t} Z^\pm(t) + 2S(t)\cdot Z^\pm(t) + \left\langle \nabla K_E(c(t)), \vec{Y}^\pm(t)\right\rangle = 0, \tag{12.28}$$

with boundary condition $\lim_{t\to-\infty} Z^\pm(t) = 0$.

Thus the l.h.s. of (12.19) equals the absolute value of

$$\int_{-\infty}^t Z_g^\pm(\tau;\theta^-,\theta^+)d\tau + \frac{Z_g^\pm(t;\theta^-,\theta^+)}{S_g(t;\theta^-,\theta^+)} - \frac{\frac{\partial}{\partial\theta^\pm}\dot{Y}_g^+(t;\theta^-,\theta^+)}{\dot{Y}_g^+(t;\theta^-,\theta^+)}. \tag{12.29}$$

We now claim that

$$\int_{-\infty}^{\infty} \left|Z_g^\pm(t;\theta^-,\theta^+)\right| dt < C \qquad \text{for } g \in G_l, (\theta^-,\theta^+) \in U_l \tag{12.30}$$

and

$$\frac{Z_g^\pm(t;\theta^-,\theta^+)}{S_g(t;\theta^-,\theta^+)} = o(1) \qquad \text{as } t \to \infty. \tag{12.31}$$

Inserting this into (12.29) proves (12.19), since in view of (12.23) the last term on the r.h.s. of (12.29) vanishes as $t \to \infty$.

To estimate $Z^\pm(t)$, one observes that as the unique solution of (12.28) it is explicitly given by

$$Z^\pm(t) = -\int_{-\infty}^t \exp\left(-2\int_s^t S(\tau)d\tau\right)\left\langle \nabla K_E(c(s)), \vec{Y}^\pm(s)\right\rangle ds. \tag{12.32}$$

To control the r.h.s., we split every scattering geodesic $c(t)$ into an incoming (I) and an outgoing (III) segment, and the bounded segment (II) where $c(t) \in \mathbf{G}_E$. So let $T^\mp \equiv T_g^\mp(\theta^-,\theta^+)$ be the time of entrance and leave of the interaction zone $\mathbf{G}_E$, respectively.

We now control the quantities occurring in (12.28) for the time intervals I-III, beginning with the gradient $|\nabla K_E(c(t))|$ of the Gaussian curvature.

By assumption, the potential V meets the set of equations (2.35) which implies in particular that asymptotically K_E converges with all derivatives to the Gaussian curvature (3.17) of a single Coulomb potential of charge Z_∞. Thus there is a $k_1 > 0$ such that for all scattering data $(\theta^-, \theta^+) \in U_l$, $g \in G_l$ and for $t \leq T_g^-(\theta^-, \theta^+)$ resp. for $t \geq T_g^+(\theta^-, \theta^+)$

$$|\nabla K_E(c_g(t;\theta^-, \theta^+))| \leq k_1 \left\langle t - T_g^\pm(\theta^-, \theta^+)\right\rangle^{-4} \tag{12.33}$$

with $\langle x\rangle := (x^2+1)^{1/2}$.

By compactness of $\mathbf{G}_E$, we can choose k_1 so that $|\nabla K_E(c_g(t;\theta^-, \theta^+))| \leq k_1$ for $T^- \leq t \leq T^+$ and all scattering data.

As the next quantity we estimate $S(t)$. We already know that $S(t) \geq 0$. But since the Gaussian curvature $K_E \leq 0$, and $K_E < 0$ for points on the geodesic with non-vanishing Euclidean curvature, there exists a strictly invariant cone field on $\mathbf{U}_E$ in the sense of Def. 5.6. By arguments analogous to those applied in the proof of Prop. 5.7, there exist $s_l > 0$, $T > 0$ such that for all scattering data $g \in G_l$ $(\theta^-, \theta^+) \in U_l$,

$$S_g(t;\theta^-, \theta^+) \geq s_l \qquad \text{for } T_g^-(\theta^-, \theta^+) + T \leq t \leq T_g^+(\theta^-, \theta^+).$$

This implies in case III, *i.e.* for $t \geq T_g^+(\theta^-, \theta^+)$,

$$S_g(t;\theta^-, \theta^+) \geq (t - T_g^+(\theta^-, \theta^+) + s_l^{-1})^{-1}. \tag{12.34}$$

Now we handle $Y^\pm(t)$. We write down the estimates for $Y^+(t)$. The estimates for $Y^-(t)$ follow from symmetry under time reversal.

Eqs. (12.22), (12.23) imply $0 \leq \dot{Y}^+(t) \leq 1$ so that

$$0 \leq Y_g^+(t;\theta^-, \theta^+) \leq s_l^{-1} + t - T^+(\theta^-, \theta^+) \qquad \text{for } t \geq T^+(\theta^-, \theta^+), \tag{12.35}$$

using $S(T^+) = \dot{Y}^+(T^+)/Y^+(T^+) \geq s_l$.

In case II we have

$$0 \leq Y^+(t) \leq k_2 \exp(s_l(t - T^+)) \qquad \text{for } T^- \leq t \leq T^+,$$

with $k_2 := \exp(s_l T)/s_l$.

Finally in case I we only know that

$$0 \leq Y^+(t) \leq Y^+(T^-) \leq k_2 \exp(s_l(T^- - T^+)) \qquad \text{for } t \leq T^-. \tag{12.36}$$

Summing up, we found

$$\left|\left\langle \nabla K_E(c(s)), \vec{Y}^\pm(s)\right\rangle\right| \leq \begin{cases} k_1 k_2 \frac{\exp(s_l(T^- - T^+)}{\langle s - T^-\rangle^4} & \text{for } s \leq T^- \\ k_1 k_2 \exp(s_l(t - T^+) & \text{for } T^- < s \leq T^+ \\ k_3 \left\langle s - T^+\right\rangle^{-3} & \text{for } s > T^+ \end{cases} \tag{12.37}$$

where k_3 depends only on s_l.

With these informations at hand, we can bound $Z^+(t)$ using (12.32). Since $S(t) \geq 0$ for $t \in \mathbb{R}$, we find in case I

$$\begin{aligned} \left|Z^+(t)\right| &\leq \int_{-\infty}^{t} \left|\left\langle \nabla K_E(c(\tau)), \vec{Y}^+(\tau)\right\rangle\right| d\tau \\ &\leq k_1 k_2 \exp(s_l(T^- - T^+)) \int_{-\infty}^{t} \left\langle \tau - T^- \right\rangle^{-4} d\tau, \end{aligned} \tag{12.38}$$

using (12.37).

In case II we obtain from (12.37)

$$\begin{aligned} \left|Z^+(t)\right| &\leq k_1 k_2 \cdot \left(\exp(s_l(T^- - T^+)) \int_{-\infty}^{0} \langle t\rangle^{-4} dt + \frac{1}{s_l}\exp(s_l(t - T^+))\right) \\ &\leq k_4 \exp(s_l(t - T^+)) \end{aligned} \tag{12.39}$$

with $k_4 := k_1 k_2 (\int_{-\infty}^{0} \langle t\rangle^{-4} dt + 1/s_l)$.

Finally, we shall show that in case III, $Z^+(t)$ converges to zero sufficiently fast as $t - T^+ \to \infty$, uniformly for all scattering data $(\theta^-, \theta^+) \in U_l$, $g \in G_l$.

Using (12.34) one finds for $s \leq t$, $t \geq T^+$

$$\int_{s}^{t} S(\tau) d\tau \geq \ln\left(\frac{t - T^+ + s_l^{-1}}{\max\{s, T^+\} - T^+ + s_l^{-1}}\right), \tag{12.40}$$

yielding

$$\exp\left(-2\int_{s}^{t} S(\tau) d\tau\right) \leq \left(\frac{\max\{s, T^+\} - T^+ + s_l^{-1}}{t - T^+ + s_l^{-1}}\right)^{2-\varepsilon} \tag{12.41}$$

for any $\varepsilon \geq 0$.

Inserting (12.41) into (12.32) and using (12.37) gives

$$\left|Z^+(t)\right| \leq k_5 \left\langle t - T^+ \right\rangle^{-2+\varepsilon}, \qquad t \geq T^+. \tag{12.42}$$

where k_5 depends only on $\varepsilon > 0$, k_3, k_4 and s_l.

Thus (12.30) follows from the estimates (12.38), (12.39) and (12.42) for the cases I,II and III, whereas (12.31) follows from (12.42) and (12.34).

The proof of (12.19) for the case $\partial/\partial\theta^-$ is similar in spirit (but not in the details) to the case above and is left to the reader.

To show existence of Lipschitz constants for the derivatives

$$\frac{\partial^{n_1+n_2}}{(\partial\theta^+)^{n_1}(\partial\theta^-)^{n_2}} \left(\frac{d\varphi_{E,\theta^-}}{dL}\right)^{-1} (L_g(E, \theta^-, \theta^+)),$$

one considers the $\theta^\pm$ derivatives of (12.28) which again lead to first order linear inhomogeneous equations parametrized by the scattering data.

Assuming (2.35), the falloff of the inhomogeneity for $t \to \pm\infty$ has the same type of bounds, with different constants. □

Remark 12.4 Under the conditions of Thm. 12.3, the ratio

$$\lim_{\theta^+ \to \theta^-} \frac{\frac{d\sigma}{d\theta^+}(E,\theta^-,\theta^+)}{\frac{d\sigma}{d\theta^+}(Z_\infty)(E,\theta^-,\theta^+)} = 1, \tag{12.43}$$

if the asymptotic charge $Z_\infty > 0$. Here the differential cross section in the denominator is the Rutherford cross section (12.1) for charge Z_∞.

The reason for (12.43) is the following: For $\theta^+ \approx \theta^-$ the main contribution to the differential cross section comes from a single orbit with large impact parameter, the other contributions being uniformly bounded in (θ^-,θ^+).

Then $|\vec{q}|$ is large throughout the orbit and, by (2.4), the additional force term $-\nabla(V(\vec{q}) + Z_\infty/|\vec{q}|)$ acts as a perturbation compared to the Coulombic term $\nabla Z_\infty/|\vec{q}|$.

13. Concluding Remarks

As we have seen in the previous analysis, the scattering theory considered has one nice feature: It is universal in the high-energy regime, in the sense that many qualitative properties are independent of the detailed form of the Coulombic potential.

Although the Coulomb singularity is basic for physics, there is a whole list of limitations of our model

1. All charges are assumed to be positive;
2. In reality, molecular potentials are time-dependent even if on disregards the interaction with the scattering particle;
3. The motion should take place in three dimensional space instead of a plane;
4. The scattering process should be described by Schrödinger's equation instead of Newton's law.

Clearly, an answer to *all* these problems is totally unfeasible, since this would amount to the formulation of a quantum theory of the full scattering problem, based on a geometrical description of the classical process.

Rather, one should try to give partial answers to the *single* questions, which are to our opinion approximately ordered by increasing difficulty.

1) The inclusion of repelling singularities ($Z_l < 0$) should be possible in the high-energy regime.

More care is necessary in that case to apply the covering construction since the Jacobi metric degenerates at those points $\vec{q}$ where $V(\vec{q}) = E$.

There is no hope to find a general description of the dynamics unless the motion is unstable everywhere.

At first sight this may seem to be in doubt for the case of repelling singularities, and in fact one should find *low*-energy stable orbits even for some potentials of the purely Coulombic type

$$V(\vec{q}) = -\sum_{l=1}^{n} \frac{Z_l}{|\vec{q} - \vec{s}_l|}, \qquad Z_l < 0.$$

Geometrically speaking, the Gaussian curvature K_E of repelling Coulomb potentials is larger than zero, as can be inferred by inspection of eq. (3.17). And positive curvature tends to collimate geodesics, whereas negative curvature leads to instability.

The clue to the understanding of the repelling case is the observation that large particle energies lead to short 'focal lengths' of the Coulombic singularities. The system becomes unstable if the focal lengths become shorter than the mutual distances of the nuclei.

Nevertheless, the inclusion of repelling singularities makes the motion less universal, because of the *shadowing* phenomenon.

Take a sum of $n = 3$ repelling Coulombic potentials. If the positions $\vec{s}_1, \vec{s}_2$ and $\vec{s}_3$ of the nuclei are arranged to be on one line, then there are only two bounded orbits (bouncing back and forth between the middle nucleus and the left resp. right nucleus).

If the three nuclei are in general position, then for large energies there exists a Cantor structure of closed orbits. Similar phenomena have been analysed in the case of smooth potential wells.

Thus we conjecture that high energy motion becomes unstable, even if one includes repelling Coulomb singularities in general position.

2) Molecules rotate and vibrate. So the positions $\vec{s}_l$ of the nuclei are periodic (or quasiperiodic or even more general) functions of time, and we should describe the scattering process by a time-dependent Hamiltonian function $H(\vec{q}, \vec{p}, t) = \frac{1}{2}\vec{p}^2 + V(\vec{q}, t)$. In the simplest case of a uniform rotation one can even get rid of the explicit time dependence.

We expect much of our analysis to carry over to this cases because of *structural stability* of our system. That is, the phase portrait should retain its qualitative form under small perturbations; more precisely, there should exist a homeomorphism from the energy shell Σ_E to the energy shell of the perturbed system carrying oriented orbits to oriented orbits.

In the case of Anosov flows the proof of structural stability is based on the existence of a hyperbolic structure on the underlying manifold.

In our case we only have a hyperbolic structure on the measure zero set of bounded orbits. This makes the sought-for homeomorphism highly non-unique. Indeed, in addition to the usual deformations in the flow direction, there are many other self-conjugacies.

For the case of a uniform rotation the relevant parameter is the ratio between the speed of the scattering electron and the speed of the nuclei. For large values of that parameter we expect to reach an 'adiabatic regime' where the phase portrait of the time dependent system in the corotating frame becomes conjugate to the one of the system considered in this book.

3) The generalization to scattering in three space dimensions is a difficult and challenging problem.

It is possible to regularize the Coulomb singularity in three dimensions, too. Indeed there exist different methods, as the one by Kustaanheimo-Stiefel, see Stiefel and Scheifele [42], and the one by Moser [33].

But there are important dynamical features which lead to a non-universal behaviour of the three-dimensional motion, even in the regime of large energies.

As an example, consider the sum of $n = 3$ attracting purely Coulombic potentials, with the positions $\vec{s}_1$, $\vec{s}_2$ and $\vec{s}_3$ of the nuclei arranged on a line. By axial symmetry, the closed orbits come in one-parameter families. The force $-\nabla V$ is tangent to any plane containing the axis. So we know from our analysis of the two-dimensional case that there are many nondegenerate one-parameter families of closed orbits.

For every such closed orbit, *two* of the three pairs of Lyapunov exponents are equal to one, which implies neutral stability of the families of closed orbits. Therefore a very weak perturbation can destroy these families so that from the topological point of view the system is not structurally stable.

Another way to see the problem is to consider the sectional curvature of the Jacobi metric (which is a generalization of the Gaussian curvature for $d \geq 3$). The sectional curvature of a Coulomb potential is not negative definite; instead it becomes positive for tangent planes perpendicular to ∇V. This is the geometric reason for the neutral stability of the above families of orbits.

So, unlike in this book, an investigation of the three-dimensional scattering problem should not be based on a topological approach. Instead, one should try to obtain results of measure theoretical nature.

4) The range of applications of our results is not limited to molecular scattering. Alternatively, one may interpret the model as a description of a fast test particle in the field of celestial bodies, whose positions change only slowly in time. Then, of course, a classical description is fully justified.

In [24] we analyse some questions concerning the related quantum mechanical problem.

In particular, for the two centre problem we relate the positions of the resonances to the length and Lyapunov exponent of the single closed orbit (by showing the validity of Bohr-Sommerfeld type formulae).

For $n \geq 3$, the fractal dimension of the bounded orbits determines an upper bound for the number of resonances within a small window near the real energy axis.

Other questions are being discussed now in the physics-oriented literature on irregular scattering, such as: Statistics of resonances, Ericson fluctuations, or enhanced back scattering. To us, it seems to be hard to obtain rigorous mathematical results on those problems.

Except from these possible generalisations of the model, there is a number of questions concerning the model which were left open:

1. Structural stability of the system;
2. Dependence of the topological entropy on the energy;
3. The meaning of the braid group action;

4. The 'prime number theorem' for the periods of the closed orbits;
5. Inverse scattering.

1) We already mentioned structural stability in point 2) above. In formulating a precise result, one needs some metric on the set of Coulombic potentials, in order to be able to define what a 'small perturbation' is.

2) In [21], Katok, Knieper, Pollicott and Weiss proved the astonishing statement that a smooth perturbation of a smooth Anosov flow leads to a smooth variation of the topological entropy.

This result is astonishing in two ways. Firstly, the foliation by the stable and unstable manifolds of smooth Anosov flows is only Hölder continuous in general. Secondly, for general flows, the topological entropy need not even vary continuously under smooth perturbations.

We conjecture a smooth variation of the topological entropy with the energy, for E large, and for all positive E if the potential is logarithmic superharmonic.

In [46], Troll analysed a piecewise linear map depending on a parameter, which serves as a model for chaotic scattering. He proved that the topological entropy of the map, considered as a function of the parameter, has the form of a devil's staircase.

For low energies $E > 0$, the topological entropy of the flow Φ_E^t could show a similar dependence on E for Coulombic potentials with a rich structure of bifurcations.

3) In Remark 6.12 we described a natural action of the braid group on the set of bounded orbits.

The existence of such a group action should not come as a surprise since our configuration space is the plane, with n points deleted.

A more detailed analysis of that group action may be of some interest, (*e.g.*, for $n \geq 3$ there is a closed orbit encircling all nuclei which is invariant under all group elements).

More important, however, is the question of whether or not this 'group symmetry' has any influence on measurable quantities, or whether it is of any use in a Gutzwiller type Ansatz for quantum mechanical scattering.

4) In Chap. 8 we showed a theorem of 'prime number type' which described the growth rate of the number of closed *geodesics* with their minimal length. We were not able to derive a similar theorem for the number of closed *orbits* of the flow Φ_E^t, since we had no tight control on the time reparametrization connected with our covering construction (see Remark 8.6). Closely related is the question of whether or not the flow, restricted to the bounded orbits is mixing w.r.t. a suitable measure (it is certainly ergodic). Of course we expect mixing, since otherwise the bounded orbits would have to show a kind of synchronisation. This synchronisation might be excluded by perturbation theory around orbits in the high-energy regime. A formal calculation (neglecting error bounds) then leads to a nonvanishing desynchronization.

5) For spherically symmetric potentials it is relatively easy (under some additional assumptions) to recover the form of the potential from the differential cross section (see, e.g., Landau and Lifschitz [30], §18).

Clearly it is not possible to recover the form of the Coulombic potential from the differential cross section alone (since that quantity is invariant under translations)

Instead of differential cross section, one could analyse the deflection function, *i.e.* the dependence of the final angle on the initial data.

The question of inverse scattering has a geometric touch, since it can be reformulated as the problem of how to recover curvature from the asymptotics of the geodesics, using the Jacobi metric.

References

[1]Abraham, R., Marsden, J.E.: Foundations of Mechanics. Reading: Benjamin 1978

[2]Anosov, D.: Geodesic flows on closed Riemannian manifolds with negative curvature. Proc. Steklov Inst., Vol. 90 (Amer. Math. Soc. translation) 1969

[3]Behnke, H., Sommer, F.: Theorie der analytischen Funktionen einer komplexen Veränderlichen. Die Grundlehren der mathematischen Wissenschaften in Einzeldarstellungen, Vol. 77. Berlin, Heidelberg, New York: Springer 1962

[4]Birman, J.S.: Braids, Links, and Mapping Class Group. Annals of Mathematics Studies. Princeton: Princeton University Press 1974

[5]Bolotin, S.V.: Nonintegrability of the n-centre problem for $n > 2$. Vestnik Mosk. Gos. Univers., ser. math. mekh. **46** (1982), No. 4

[6]Bowen, R.: Periodic Orbits for Hyperbolic Flows. Amer. J. Math. **94**, 1–30 (1972)

[7]Combes, J.M., Duclos, P., Klein, M., Seiler, R.: The Shape Resonance. Commun. Math. Phys. **110**, 215–236 (1987)

[8]Eckhardt, B.: Irregular Scattering. Physica D **33**, 89–98 (1988)

[9]Eckhardt, B., Jung, Ch.: Regular and irregular potential scattering. J. Phys. A: Math. Gen. **19** L829–L833 (1986)

[10]Falconer, K.J.: The geometry of fractal sets. Cambridge: Cambridge University Press 1986

[11]Farkas, H.M., Kra, I.: Riemann surfaces. Graduate Texts in Mathematics, Vol. 71. Berlin, Heidelberg, New York: Springer 1977

[12]Fomenko, A.T.: Integrability and Nonintegrability in Geometry and Mechanics. Dordrecht, Boston, London: Kluwer 1988

[13]Franz, W.: Topologie I. Berlin, New York: Walter de Gruyter 1973

[14]Gutzwiller, M.: Mild Chaos. In: Chaotic Behavior in Quantum Systems. Ed.: G. Casati. New York and London: Plenum Press 1985

[15]Hadamard, J: Sur les lignes géodésiques des surfaces à courbures opposées. Procès-Verbaux Soc. Sci. Phys. Natur. Bordeaux, 4. mars 1897.

[16]Hadamard, J: Les surfaces à courbures opposées et leur lignes géodesiques. Journ. Math. 5e série, t. 4, 27–73 (1898); Œvres, Tome II

[17]Helffer, B., Sjöstrand, J: Resonances en limite semi-classique. Bulletin de la S.M.F., memoire No. 24/25, **114** (1986)

[18]Hirsch, M.W.: Differential Topology. Graduate Texts in Mathematics, Vol. 33. Berlin, Heidelberg, New York: Springer 1988

[19]Hunziker, W.: Scattering in Classical Mechanics. In: Scattering Theory in Mathematical Physics. J.A. La Vita and J.-P. Marchand, Eds., Dordrecht: Reidel 1974

[20]Jung, Ch., Tél, T.: Dimension and escape rate of chaotic scattering from classical and semiclassical cross section data. *preprint* (1991)

[21]Katok, A., Knieper, G., Pollicott, M., Weiss, H.: Differentiability and Analyticity of Topological Entropy for Anosov and Geodesic Flows. Inventiones mathematicae **98**, 581–597 (1989)

[22]Klein, M.: On the Absence of Resonances for Schrödinger Operators with Non-Trapping Potentials in the Classical Limit. Commun. Math. Phys. **106**, 485–494 (1986)

[23]Klein, M.: On the Mathematical Theory of Predissociation. Annals of Physics **178** 48–73 (1987)

[24]Klein, M., Knauf, A.: *In preparation*

[25]Klingenberg, W.: Eine Vorlesung über Differentialgeometrie. Berlin, Heidelberg, New York: Springer 1977

[26]Klingenberg, W.: Riemannian Geometry. Studies in Mathematics 1; Berlin, New York: De Gruyter 1982

[27]Knauf, A.: Ergodic and Topological Properties of Coulombic Periodic Potentials. Commun. Math. Phys. **110**, 89–112 (1987)

[28]Knauf, A.: Coulombic Periodic Potentials: The Quantum Case. Annals of Physics **191**, 205–240 (1989)

[29]Knauf, A.: Closed orbits and converse *KAM* theory. Nonlinearity **3**, 961–973 (1990)

[30]Landau, L.D., Lifschitz, E.M.: Lehrbuch der theoretischen Physik, Vol. I. Berlin: Akademie-Verlag 1966

[31]Loomis, L.H., Sternberg, S.: Advanced Calculus. Reading: Addison-Wesley 1968

[32]Milnor, J.: Morse Theory. Annals of Mathematics Studies. Princeton: Princeton University Press 1973

[33]Moser, J.: Regularization of Kepler's Problem and the Averaging Method on a Manifold. Comm. Pure Appl. Math. **23**, 609–636 (1970)

[34]Narnhofer, H.: Another Definition for Time Delay. Phys. Rev. D **22**, 2387–2390 (1980)

[35]Narnhofer, H., Thirring, W.: Canonical Scattering Transformation in Classical Mechanics. Phys. Rev. A **23**, 1688–1697 (1981)

[36]Schwartz, J.T.: Nonlinear Functional Analysis. New York: Gordon and Breach 1969

[37]Parry, W., Pollicott, M.: An analogue of the prime number theorem for closed orbits of Axiom A flows. Ann. Math. **118**, 573–591 (1983)

[38]Poincaré, H.: Œvres, Vol. 6, Paris, Gauthier-Villars 1953

[39]Simon, B.: Wave Operators for Classical Particle Scattering. Commun. Math. Phys. **23**, 37–49 (1971)

[40]Sinai, Ya. G., Ed.: Dynamical Systems II. Encyclopaedia of Mathematical Sciences, Vol. 2. Berlin, Heidelberg, New York: Springer 1989

[41]Smilansky, U.: The Classical and Quantum Theory of Chaotic Scattering. Lecture Notes Summer School on 'Quantum Chaos'. Les Houches 1989.

[42]Stiefel, E.L., Scheifele, G.: Linear and regular celestial mechanics. Grundlehren der mathematischen Wissenschaften, Vol. 174. Berlin, Heidelberg, New York: Springer 1971

[43]Tél, T.: Transient Chaos. In: Directions in Chaos, Vol. 4 Ed. Hao Bai-lin. Singapore: World Scientific 1990

[44]Thirring, W.: Lehrbuch der Mathematischen Physik 1. 2nd Ed.; Wien, New York: Springer 1988

[45]Tricot, C., Jr: Two definitions of fractional dimension. Math. Proc. Camb. Phil. Soc. **92**, 57–74 (1982)

[46]Troll, G.: How to escape a sawtooth. The Weizmann Institute of Science. Preprint

[47]Walters, P.: An Introduction to Ergodic Theory. Graduate Texts in Mathematics, Vol. 79. Berlin, Heidelberg, New York: Springer 1982

[48]Wojtkowski, M.: Invariant families of cones and Lyapunov exponents. Ergod. Th. & Dyn. Systems **5**, 145–161 (1985)

Index of Symbols

Index